CW01651971

# THE MUSHROOM IDENTIFIER

# THE MUSHROOM IDENTIFIER

DAVID PEGLER & BRIAN SPOONER

Grange
BOOKS

A QUANTUM BOOK

Published by Grange Books
An imprint of Grange Books plc
The Grange
Grange Yard
London SE1 3AG

Copyright © 1992 Quintet Publishing Limited

This edition printed 1997

All rights reserved.
This book is protected by copyright. No part of it may be reproduced, stored in a retrieval system, or transmitted in any form or by any means, without the prior permission in writing of the Publisher, nor be otherwise circulated in any form of binding or cover other than that in which it is published and without a similar condition including this condition being imposed on the subsequent publisher.

ISBN 1-85627-949-9

QUMMID

This book was produced by
Quantum Books Ltd
6 Blundell Street
London N7 9BH

Printed in China by Leefung-Asco Printers Ltd

**A number of mushrooms are poisonous and some are deadly poisonous. It is therefore vitally important that you are absolutely sure of the identity of any mushroom before eating it. It is always best to consult an expert. If there is any doubt at all about the identity of a mushroom, do not eat it.**

**The publisher and authors of this book have made every effort to ensure that the information contained in it is accurate and factually correct. Any information regarding the edibility of any mushroom is provided for guidance only. The publisher and authors can not accept legal responsibility for any errors or omissions or for any actions taken arising from the information contained in this book.**

# Contents

INTRODUCTION 6

**DIRECTORY**

ON THE GROUND IN WOODLANDS, OR ASSOCIATED WITH TREES 18

ON THE GROUND IN GRASSLANDS OR OPEN SPACES 72

ON TREES, STUMPS OR WOODY DEBRIS 93

IN WET SITUATIONS, SUCH AS BOGS AND MARSHLANDS 127

ON BURNT GROUND OR ON BURNT WOOD 130

ON DUNG OR ENRICHED SOIL 134

ON OTHER FUNGI 137

FURTHER READING 140

GLOSSARY 140

INDEX 142

CHAPTER ONE

# INTRODUCTION

# PURPOSE OF THE BOOK

This book provides the reader with an opportunity to identify many of the larger fungi which might be encountered during a walk through woods and fields. The guide provides colour photographs and detailed descriptions of 341 of the most common and more interesting of the larger fungi which are to be found both in North America and Europe.

There are probably more than 8,000 species of these fungi to be found growing in this particular geographical area, and it has therefore been necessary to make a careful selection. Many of the known species are rare and unlikely to be encountered by the casual observer. Furthermore, many closely related species can be separated only by careful and detailed examination of their minute structures using a powerful compound microscope. It should always be remembered, therefore, that the precise identification of a species is a difficult and skilled task.

# CHOICE OF SPECIES

The species chosen have been, for the most part, selected on the basis that they are likely to be found, that they are the most readily identifiable, and that they are widely distributed throughout the northern hemisphere. Most readers will be familiar with mushrooms and toadstools, but there are many other examples of the "larger fungi", such as bracket-fungi, club-fungi, tooth-fungi, puffballs, stinkhorns, jelly-fungi, cup-fungi, and flask-fungi. Examples of all these groups are represented in this book, and the choice is left to the reader as to which groups are most of interest.

# HOW TO USE THE BOOK

The book groups the species under seven different headings, each based either upon the habitat or ecological area in which the species grows or on the substrate upon which the fruitbodies may be found. For example, species may grow in woodlands or in grassland, or are perhaps confined to boggy or marshland areas. The fruitbodies may be found growing on soil or leaf litter, or on wood, on dung, on burnt ground, or even on other fungi. When you wish to identify your find, first note exactly where and under what conditions you found the fruitbody growing, and then select the relevant chapter. Within each chapter, the fungi are arranged according to their groups: for example, chanterelles, mushrooms and toadstools, bracket-fungi, and so on. To help you, accounts of each of these groups are to be found on pages 7–10. In addition to a detailed description for each species, further information is provided on the habitat, the season and the edibility or poisonous nature of the species, and reference is made to related species with which confusion may occur. Refer to the glossary for definitions of specialist botanical terms used in species descriptions.

# MAJOR GROUPS OF FUNGI

The kingdom Fungi is an enormous group of living organisms. They are quite distinct from either animals or plants, and the group is probably as large as, or larger than, either. Fungi may be found in a great variety of forms, ranging from the single cells of the yeasts, through the thread-fungi and pin-moulds on rotting food, to the Giant Puffball or the long-lived large bracket-fungi on tree-trunks. Most fungi consist basically of fine, microscopic threads, called hyphae, which live and spread within a substratum, such as soil, wood, or other plant tissues. In order to spread and multiply, fungi produce spores and form special structures, called fruitbodies, for this purpose.

Fungi are classified according to the manner in which they produce their spores and the type of fruitbody they develop. There are three main classes of fungi: the Zygomycetes, the Ascomycetes and the Basidiomycetes. The Zygomycetes are the thread-fungi and pin-moulds, and do not normally form a fruitbody which is sufficiently large to be seen with the naked eye, so they are not considered further in this account. The Ascomycetes produce their spores (usually eight) within a single "cell", called an ascus, and their fruitbodies may be open cup- or disc-shaped structures (Discomycetes), or closed, flask-shaped structures (Pyrenomycetes). The Basidiomycetes produce their spores (usually four) on short stalks (sterigmata) at the top of a club-shaped cell, called a basidium, and the fruitbodies may be mushrooms, brackets, clubs, and so on.

## CHANTERELLES

This small group includes short-lived, fleshy fungi which initially resemble mushrooms and toadstools in having a central stem and are either funnel-shaped and hollow, as in *Craterellus cornucopioides,* or are flattened on top into a cap-like structure. Unlike mushrooms, which have clearly defined, radiating gills on the underside of the cap, chanterelles have thick, branching ridges with blunt edges. The Chanterelle is one of the best-known edible fungi and is often served in restaurants.

## Mushrooms And Toadstools

Mushrooms and toadstools form the bulk of the species illustrated in this book, and are the fungi most frequently observed. They are especially frequent in woodlands during the autumn months. There is no scientific distinction between a mushroom and toadstool, and there are many exceptions to the common belief that "mushrooms" are edible species while "toadstools" are poisonous. Traditionally, both terms exist in the English language, whereas only a single term exists in most other languages.

The scientific term for these fungi is agarics (order Agaricales). They are Basidiomycetes, producing their spores on gills, which radiate out from the underside of a protective cap that is usually, although not always, supported by a centrally or laterally attached stem. They are mostly soft, fleshy fungi, growing very quickly but surviving for only a few days or even, perhaps, a few hours.

There are many families of mushrooms and toadstools, distinguished largely by differences in their microscopic structures. One of the field characters useful for identifying them is the colour of their spores when deposited in mass. Species may be white-, pink-, brown- or black-spored. Other valuable field characters include the presence or absence of protective layers, called veils, the manner of attachment of the gills to the top of the stem, and any special association with a tree or substratum.

Mushrooms and toadstools include the commercial, edible mushrooms, the deadly poisonous *Amanita* species, the brightly coloured brittle-gills, the milk-caps, oyster-caps and wax-gills. There are well over 150 genera and more than 5,000 species in the northern hemisphere.

## Boletes

The boletes (order Boletales) are related to the mushrooms and toadstools, and for the most part are soft and fleshy, with a short life-span. They usually have a fleshy cap supported by a central stem but, unlike the mushrooms, the spores are produced not on radiating gills but inside a series of closely compacted vertical tubes, each opening by a basal pore, on the underside of the cap. The ceps and their relatives are among the most highly prized of the edible fungi. Some species look like mushrooms with gills, for example *Paxillus* and *Gomphidius,* but the structure of their spores and other characteristics indicates a natural relationship with the boletes. The bracket-fungi also have tubes and pores on the underside of the cap, but are much tougher.

## Club-Fungi And Tooth-Fungi

The club-fungi (or Clavarias) and the tooth-fungi (or Hydnums) are also, for the most part, soft and fleshy fungi. They are often brightly coloured. Most species grow on the ground, although a few occur on dead wood. The club-fungi form fruitbodies which resemble erect, simple clubs that are either solitary or clustered, or the fruitbodies may be highly branched. The latter are often referred to as coral-fungi, for example *Ramaria* species. Although most species are harmless, they are rarely recommended for eating, and a few will cause serious stomach upsets. The tooth-fungi resemble both mushrooms and boletes by having a cap and stem, but the spores are produced on downwardly projecting teeth or spines. They are distinctive fungi but tend to be less common than other groups.

## Bracket-Fungi

These are a large group, of which the majority grow on living or dead wood, and they are therefore found mostly in woodland, forests and plantations. They fall into two main groups, the polypores and the stereums. Unlike the chanterelles, mushrooms, club- and tooth-fungi, the brackets tend to have a tough texture and develop slowly. They are long-lived and in some cases the fruitbodies are perennial. The polypores resemble the boletes in that the spores are produced within vertical tubes opening by pores on the underside of the cap. They probably represent the most advanced condition of all fungi, having complex systems of hyphae forming the flesh, and sometimes developing complex surface crusts for protection against the weather and insect attack. They were formerly all placed within a single family, Polyporaceae, but they are now considered to represent many families derived from a range of evolutionary lines. Some tough, mushroom-like species with "gills", for example *Lentinus* species, are now known to be more closely related to the polypores than to mushrooms.

The fruitbodies of stereums often form brackets on the trunks and branches of trees and have a tough texture, much in the same way as the polypores, yet they do not have tubes and pores on the underside of the cap but have spores produced on a simple, smooth surface. Many fruitbodies in this group do not form brackets with a projecting cap, but instead are very thin and closely appressed to the bark or exposed wood, and can even resemble a splash of paint. Many brackets are serious pathogens of forest and plantation trees, while others, such as the Dry Rot Fungus and the Cellar Fungus, can attack woodwork in the home. The field characters to look out for in these fungi include the colour of the flesh (often either white or rusty brown), the colour of the spore deposit, the texture and firmness of the flesh, the layering of the tube-layer (in perennial species), and the tree with which they are associated.

## Puffballs And Relatives

Puffballs, earthstars, bird's nest fungi and earthballs collectively have been placed in an order called Gasteromycetes (or "stomach fungi"), but many are now thought to have been derived from different lines of evolutionary descent, and they are no longer considered to constitute a natural group. They are Basidiomycetes but have evolved a fruitbody which remains unopened and encloses the spores. The spores are either released in clouds through a pore or the fruitbody breaks down due to weathering. Some relatives of the puffballs have developed protection from desiccation by growing underground as "false truffles". The true puffballs, for example *Lycoperdon, Bovista, Calvatia* and *Langermannia,* have white, fleshy fruitbodies when young but, as they ripen, develop a papery wall and produce large clouds of brown spores. In the case of the earthstars (*Geastrum* species), the spore-sac is protected by an outer layer which at maturity splits open into star-like rays. The earthballs (*Scleroderma* species) have a hard, round fruitbody which eventually breaks open to release the spores. The bird's nest fungi (*Cyathus* and *Nidularia* species) are a special case, in which the cup-like fruitbodies contain seed-like packets of spores which are ejected by the splashing action of raindrops. All puffballs are edible when the flesh is young.

## Stinkhorns

The stinkhorns and cage-fungi are remarkable Basidiomycetes which have previously been included in the order Gasteromycetes along with the puffballs. They initially form a soft, gelatinous "egg-stage", with a membranous outer layer which eventually ruptures to release either a tall, spongy stem (as in *Phallus* and *Mutinus* species) or a net-like structure (as in *Clathrus* species), that expands extremely rapidly and bears a slimy region containing the spores. The slime has an extremely strong and unpleasant smell but is attractive to flies. The spores are eaten and thereby dispersed.

## Jelly-Fungi

Jelly-fungi are well named, often resembling gelatinous blobs which may not be recognized as fungi. In dry conditions these fungi often dry down to a fine film on woody substrata, but become reconstituted in wet weather. All species grow on wood. They are usually yellow, white or blackish, and may form small pustules (eg *Dacrymyces* species), club shaped fruitbodies (eg *Calocera* species), brain-like gelatinous blobs (eg *Tremella* species) or brackets (eg *Auricularia* species). Although they are Basidiomycetes, they have unusual basidia which may be forked or divided. Species are not generally regarded as edible, although *Auricularia polytricha* and *Tremella delica* are often found in Chinese cuisine, though more for their texture than for their flavour.

## Cup-Fungi And Flask-Fungi

These belong to the Ascomycetes, the largest of all groups of fungi, with at least 8,000 species in the northern hemisphere. Most lichens, which are fungi having an association with an algal partner, are Ascomycetes. The Ascomycetes include yeasts, powdery mildews, cup-fungi, morels, true flask-fungi, false flask-fungi and truffles. The cup-fungi (or Discomycetes) are the most colourful and perhaps the most easily recognized (eg *Aleuria* and *Peziza* species). These produce a fruitbody which is a cup- or disc-shaped structure in which the inner surface is lined with a layer of spore-producing asci. Others are more elaborate, with a stalk-like base and a convoluted surface (eg *Morchella* and *Helvella* species). The flask-fungi (or Pyrenomycetes) form a rounded or bottle-shaped fruitbody which opens by a small apical pore and encloses the asci. Many Pyrenomycetes (eg *Xylaria* and *Cordyceps* species) produce numerous such structures which are embedded in a common fleshy layer (or stroma). Ascomycetes can grow on the soil, living and dead vegetation, other fungi, dung, or on animal remains. Morels and truffles are examples of Ascomycetes which are regularly collected and eaten.

# Collecting Wild Fungi

## When To Collect

Fleshy fungi tend to appear seasonally. The most productive months are those of the autumn, starting after the summer heat has passed and continuing through to the first frosts. As the autumn progresses, the collection of species growing on the ground can become more difficult as fallen leaves obscure the fruitbodies. The length of the collecting season varies according to climate. On the East Coast of North America and in the UK and the rest of Europe, there are two seasons, spring (April–May) and autumn (August–November). Elsewhere in North America seasons vary. In the Gulf States of America, species continue to fruit throughout the winter, whereas on the West Coast fruiting is inhibited only during the dry months. Because of the range of fruiting times in North America, it is not possible to be precise as to season in the individual entries. Indications of spring and autumn are a guide only.

## How To Collect

Most large fungi are easily collected because they lack deeply penetrating "root-systems". Care must always be taken, however, not to damage the roots and branches of trees or to uproot neighbouring plants. Only collect good, healthy specimens. Do not pick more fruitbodies than you really need, but at the same time ensure that you have sufficient specimens to demonstrate all the stages of development, for the fungus can change dramatically in appearance. Picking must be done carefully as many species are brittle and it is easy to leave behind the base of the fruitbody which may be an important characteristic when it comes to identification. In the field, only a few accessories are required. A sharp knife is essential, and a small hand-trowel is often useful. A notebook and pencil are also important for recording the substratum, the tree association, the colour of the fresh specimen and the smell. Remember that it takes time to record all the information necessary for accurate identification, so limit yourself to a reasonable number of species on each collecting trip. A good hand-lens (×10 magnification) is invaluable for viewing details of the surface structure, the attachment of the gills and similar features. Specimens are most easily transported in a flat-bottomed basket. It is recommended that fruitbodies of different species should be kept separate, and this may be achieved by carrying a collection of small boxes and tins with firm lids, or by wrapping collections in grease-proof (wax) paper. Plastic bags are not generally recommended, as the fruitbodies will "sweat" and are easily squashed. Keeping the species separate is vitally important if you are collecting for the dinner table.

At as early a stage as is possible, an attempt should be made to make a spore print. In the case of the Basidiomycetes, a spore print may be obtained by placing the fruitbody with the fertile (lower) surface facing downwards onto a sheet of white paper or a glass-slide, and leaving overnight. The specimen should be covered, or preferably placed in a tin or under a glass jar, to prevent desiccation. In the case of the mushrooms and toadstools, first remove the stem and place the cap onto the paper. Spore prints for Ascomycetes, particularly the large Discomycetes, may be obtained in a similar fashion, but the glass-slide should be placed above the fertile (upper) layer as close to the fungus as possible. Spore prints are air-dried and can be preserved with the dried specimens.

Beginners generally need guidance from an expert, and it is useful to attend "fungus forays" which are arranged each season by local and national mycological societies and natural history societies.

**Remember, before foraging, consider the land-owner and always make sure you have permission.**

# Examination And Structure Of Larger Fungi

## Examination And Preservation Of Fresh Fruitbodies

Specimens of fruitbodies should be examined as soon as possible, for most fleshy fungi have a high water content and will perish very quickly. It is always a good idea to make a colour sketch of each find, recording not only the shape and colour of the specimens but also the range encountered. Always break open the flesh to discover whether or not a colour change takes place. Taste and smell are important characters, but beginners must be very sure of their knowledge before attempting to taste any species. **If in doubt then leave well alone; some toadstools are deadly poisonous even in small amounts.** Once the overall form of the fruitbody has been carefully examined and recorded, the next stage is to cut a specimen in half vertically. This will reveal the attachment of the gills or tubes to the stem apex, and whether the stem is hollow or solid.

Many collectors wish to preserve their specimens and this can be done by drying the fruitbodies. Fruitbodies are best dried rapidly over a flow of warm air. When the heat is too low, it will only succeed in hatching insect eggs which may be deposited in the flesh of the specimen; the resultant grubs will eat the remaining fungus tissue overnight. Too high a temperature will lead to cooking, leaving a charred specimen. About 104°F/40°C is the recommended temperature, and this can be achieved by using a fan heater or a 60W lamp. Once the specimens are thoroughly dried they may be stored in envelope-type labelled packets, recording the identity of the species, the habitat, the locality, the collector and the date. Care must be taken in storage to avoid damp conditions for the specimens can quickly become mouldy. The addition of an insect repellant is strongly recommended.

## STRUCTURE OF BASIDIOMYCETES

### MUSHROOM TYPE

A mushroom is the spore-producing fruitbody of a Basidiomycete, and may be compared to the flower of a plant. The "plant" of the Basidiomycete is the network of hyphae (or mycelium) which spreads through the substratum upon which the fungus grows. The fruitbody may live only for a few days or hours, but the hidden mycelium can live for months, years, or even hundreds of years. The fruitbody acts as an umbrella or parasol, and comprises a cap (or pileus) which protects the underlying spore-producing layers from the rain and sun. The spores are produced on the surfaces of a series of radiating gills (or lamellae), which hang vertically from the cap. A valuable field character is the mode of attachment of the gills to the apex of the stem: adnate (attached by the full gill width, eg *Panaeolus*), adnexed (gills only reach the extreme stem apex, eg *Collybia*), free (gills do not reach the stem, eg *Agaricus*), sinuate (gills curve upwards just before reaching the stem, eg *Tricholoma*) or decurrent (gills run down the stem, eg *Clitocybe*). The cap may be raised above ground level or away from the woody substratum by a stem (or stripe), which is either centrally, excentrically or laterally attached to the cap. The fruitbodies vary in size from being only a fraction of an inch or a few millimetres high to, in exceptional cases, being more than one yard or one metre in diameter.

A fruitbody originates as a tiny button which grows and expands by a combination of cell-division and the rapid intake of water. The "flesh" is made up of hyphae, which appear as long chains of cells. When most structures have been formed within the button-stage, there is a rapid intake of water so that the fruitbody expands; it may appear overnight. This is known as "mushroom growth" and in the past

### STRUCTURE OF MUSHROOM

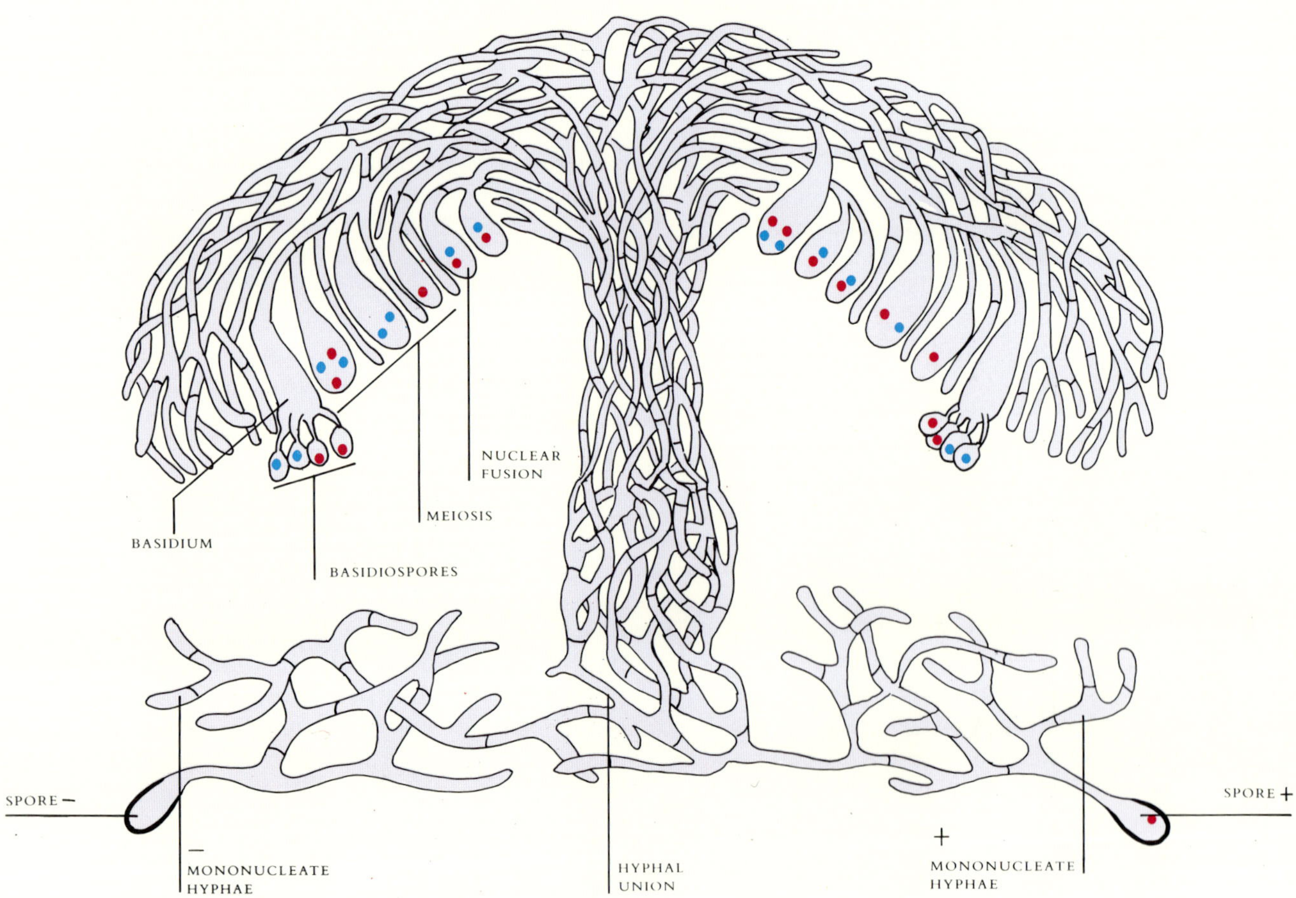

led to much speculation in folklore on the mystery of their sudden appearance. The fruitbodies of the chanterelles, many of the club-fungi and some of the tooth-fungi also have this form of growth.

Additional layers, called veils, are present in some mushrooms, especially *Amanita* species. These layers protect the young mushroom. There may be a universal veil which envelopes the entire button stage, but on expansion breaks down to leave scales over the cap surface and sometimes a sac-like volva at the stem base. In addition there may be a partial veil, which covers the young gills, but ruptures on expansion to leave a ring (or annulus) attached to the upper region of the stem.

The spores (or basidiospores) are produced by a single layer of club-shaped, single-celled basidia covering the gill surfaces. The spore is a single, undivided cell, which has a form which is characteristic of each species. Most are about 0.01 mm long, and may have a thin, colourless wall or a thick, pigmented wall. There is a small structure at the base of the spore by which it was attached to the basidium, and in some instances there is a germ-pore at the apex. This is a thinning of the spore wall through which it is able to germinate. The basidium is remarkable in that within it both conjugation and division of nuclei take place. It then produces small, apical projections (or sterigmata) of which there are typically four, and on each of which a basidiospore is formed. When the spores are ripe they are actively shot off and released into the air currents between and beyond the vertical gills, and are thereby dispersed by the wind. In the case of the ink-caps (*Coprinus* species) the mature gills digest themselves dissolving into an inky fluid, which falls to the ground taking the spores with it.

### BRACKET-FUNGUS TYPE

The fruitbodies of bracket-fungi are generally much firmer in texture, grow more slowly, and live longer than those of mushrooms. This is because their mode of growth is different. Whereas the hyphae that compose the flesh of a mushroom are thin-walled and inflate through rapid water intake to cause expansion, the hyphae of a bracket do not inflate and frequently have a thickened wall. Growth occurs by the production of more hyphae which tightly intertwine and result in a firm, rigid structure, sometimes living for several years (eg *Ganoderma* species). There may also be different types of hyphae present, some actively growing and branching, while others form a rigid skeleton to give strength to the fruitbody. The fruitbodies of the polypores, stereums and some tooth-fungi grow in this manner.

## STRUCTURE OF BASIDIOMYCETES

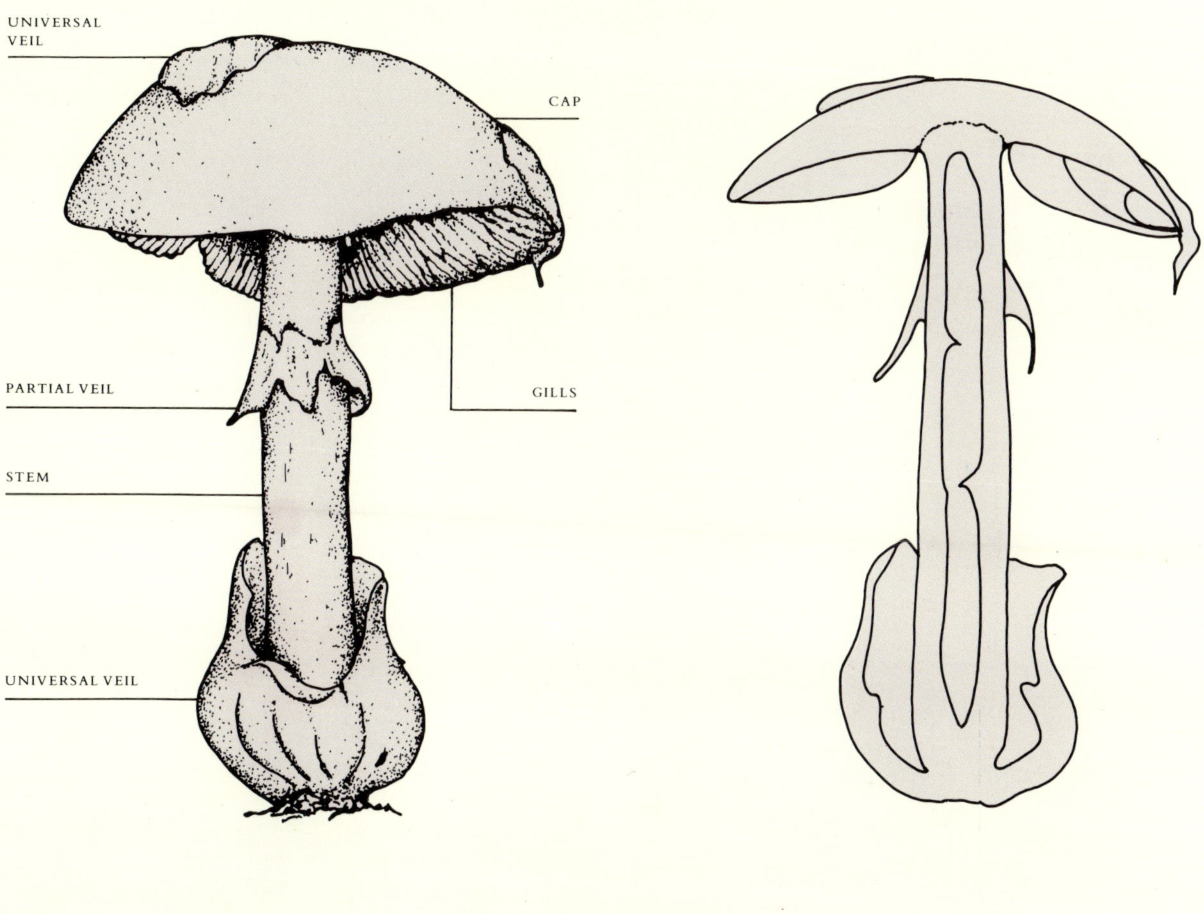

## STRUCTURE OF ASCOMYCETES

### DISCOMYCETE TYPE

The cup-fungi produce a fruitbody which is usually in the form of a cup or a disc, although some species, such as the morels (*Morchella* species) and the underground truffles (*Tuber* species) have much more convoluted fruitbodies. Unlike the Basidiomycete fruitbody the fertile, spore-producing surface (or hymenium) covers the upper surface rather than the lower surface. The rest of the fruitbody is composed of tightly interwoven, thin- or sometimes thick-walled hyphae, although there may be protective hairs and other structures on the outer surface. The fertile surface consists of a single, erect layer of spore-producing asci, which are usually cylindrical cells in which the spores (or ascospores), often eight, are formed. When the spores are ripe they are in most cases forcibly shot out through an opening at the top of the ascus. Examples of this type of fruitbody may be found in species of *Aleuria, Peziza,* and *Scutellinia*.

### PYRENOMYCETE TYPE

The spores are produced by a layer of asci, as in the Discomycetes, but this layer is contained within an almost enclosed fruitbody, often flask- or bottle-shaped, which opens by a small pore at the top of a neck. Each fruitbody is called a "perithecium", but in many cases these perithecia are aggregated into a common fleshy structure (or stroma) to form a compound fruitbody. Examples of such fruitbodies may be found in *Daldinia, Xylaria* and *Cordyceps*.

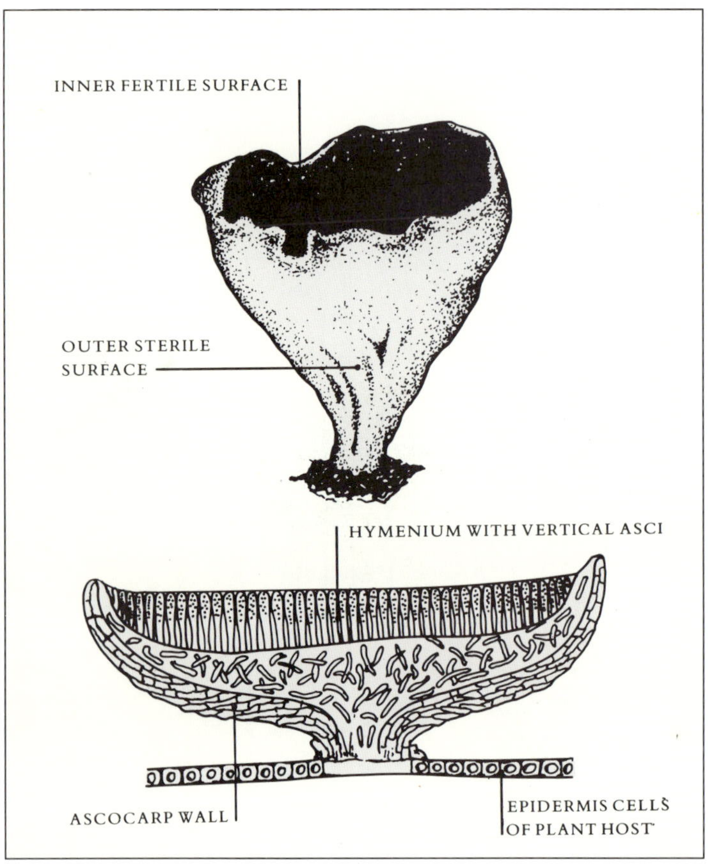

## HABITATS

When hunting for larger fungi, it is important to realize that the species you are looking for are very much dependent upon the habitat in which they live. Fungi grow where they do because of a complex pattern of interactions with the climate, the soil, plants and animals. Equally, many plants, especially trees, are dependent upon the fungi with which they grow in association. Knowing the habitat, and recognizing the tree species, will help considerably in the task of identification of the fungi. For this reason, the fungi treated in this book are considered in chapters relating to seven different habitat categories.

### 1 ON THE GROUND IN WOODLANDS, OR ASSOCIATED WITH TREES

In this habitat mycorrhizal fungi predominate. These fungi live in special association with a particular species of tree. This is why in temperate forests it is common, during the autumn months, to find large troops of fleshy mushrooms and toadstools associated with certain trees. This is especially true of genera such as *Cortinarius, Russula* and *Lactarius*. In a mycorrhizal association, the hyphae of the fungus grow in close contact with the roots of a tree. This is a symbiotic relationship in which both partners apparently gain some nutritional advantage.

### 2 ON THE GROUND IN GRASSLAND OR OPEN SPACES

Fungi which flourish in these situations must be able to tolerate wide extremes of microclimate as the environment is much more subject to fluctuation than that at the forest floor. The species do not grow in mycorrhizal association with flowering plants, and may occur throughout the year. Some species, especially Wax Caps, are brightly coloured and are conspicuous in lawns and pastures.

### 3 GROWING ON TREES, STUMPS OR WOODY DEBRIS

Species living on woody material, either living or dead, secrete enzymes which enable the breakdown of cellulose, lignin, or both. Those growing as parasites tend to be host specific. They may coexist with the living tree for many years, or may kill it more rapidly and continue to live on the dead tissue. Other fungi are able to colonize dead wood, causing a slow rot. Rotten wood, which tends to retain moisture, itself provides a suitable habitat for yet other fungi and is one of the most important habitats in temperate regions of the world.

### 4 GROWING IN WET SITUATIONS, SUCH AS BOGS AND MARSHLAND

A number of species of fungi are adapted to environments such as bogs and marshes, which are continuously waterlogged. These habitats present a special challenge to fungi. Oxygen is often limited, and fungi must therefore be adapted to tolerate anaerobic conditions. Mosses are often the dominant green plants in wet habitats, and some fungi of genera such as *Hypholoma* and *Galerina* grow in association with them. Other species, mostly of microfungi, are able to live in open, fresh or brackish water, but these species are beyond the scope of this book.

### 5 GROWING ON BURNT GROUND OR ON BURNT WOOD

This habitat has particular richness because of the release of mineral nutrients results from burning, but offers little in the way of cellulose and lignin. Charred sticks may provide both organic and inorganic nutrients, and may also retain water well. They therefore provide a particularly good substrate for fungi. Newly burnt sites are sterile, and fungi which are able to colonize at this stage are free from competition. A succession of species occurs as the amount of organic material increases.

### 6 GROWING ON DUNG OR ENRICHED SOIL

Species which fruit on dung have the advantage of a moist medium, rich in nutrients. Dung breaks down quite rapidly, and species have only a short time in which to fruit. Often, fungi are restricted to the dung of particular animals such as horse or rabbit, as the spores are ingested by the animal with its food and pass through the gut to be excreted in the dung. The animal therefore provides a method of dispersal for the spores of fungus.

Soil enriched by compost or manure offers a more stable habitat than dung, and contains nutrients in a readily assimilable form for a longer period. Many species, such as *Peziza vesiculosa,* are specially adapted to such conditions.

### 7 GROWING ON OTHER FUNGI

Conditions which are favourable for the host fungus are also favourable for fungi which grow on it. These utilize the fungal metabolites but, as with fungi on dung, the habitat is ephemeral. The colonizing fungus must complete its spore production before the host fruitbody decomposes.

# EDIBLE FUNGI, POISONOUS FUNGI, CULTIVATION OF FUNGI

## EDIBLE FUNGI

Fleshy fungi have been widely eaten by humans since ancient times, and in many countries, especially in eastern Europe, the gathering of wild mushrooms is carried out on a commercial scale. Fungi add a variety of flavours and textures to any cuisine, and they generally contain more digestible protein than any green vegetable. In addition they are high in carbohydrates, fibre and vitamins, and low in fats. Wild edible mushrooms should be picked carefully, avoiding any old or insect-infested specimens. It is recommended not to mix species in any one dish, and to be absolutely certain of the identification. The most favoured mushrooms include the chanterelles, the Cep, the Giant Puffball, the Field Mushroom, the Shaggy Ink-cap, the Parasol Mushroom, the blewits, the Cauliflower Fungus, the Wood Hedgehog, and the morels.

**REMEMBER**

- *First, learn to recognize the deadly species, especially the Death Cap*
- *Make certain of identification, and do not mix species.*
- *Use only fresh and clean specimens, and ensure that they are properly cooked.*
- *If you are trying a known edible species new to you, only sample a small quantity.*

## Poisonous Fungi

Some wild fungi are deadly poisonous, even in small amounts. On returning from a collecting trip, always remember to wash your hands. There are a number of different types of fungus poisoning, and the symptoms of poisoning can therefore vary.

By far the most dangerous form of poisoning is that caused by the Death Cap (*Amanita phalloides*) and its relatives, attacking the liver and often proving fatal. Symptoms include abdominal pains, nausea, vomiting and intense thirst starting 8–15 hours after ingestion, followed by a temporary recovery period. In the case of orellanin poisoning, caused by the brown-spored *Cortinarius speciossimus* and allies, the onset of symptoms does not take place until 36–48 hours after ingestion. The Turban Fungus (*Gyromitra esculenta*) is another fungus which can prove fatal, and is sometimes confused with the edible morels.

Other species attack the blood cells when eaten raw, resulting in anaemia. Such species include The Blusher (*Amanita rubescens*), and the Grisette (*A. vaginata*). In the case of the Common Ink-cap (*Coprinus atramentarius*), this otherwise edible mushroom will react violently, within 10 minutes, with any alcohol consumed during the same meal. A number of toadstools, for example (*Amanita muscaria* and small, white *Clitocybe* species, contain a substance called muscarine which causes excessive perspiration, together with nausea and vomiting.

Some species affect the central nervous system causing hallucinations and sometimes leading to coma. In the case of muscimol poisoning, also caused by the Fly Agaric (*Amanita muscaria*) and by others such as The Panther (*A. pantherina*), the symptoms consist mainly of drowsiness but can be more serious. Some of the *Psilocybe* species, on the other hand, cause visual hallucinations within 20 minutes of ingestion. Such mushrooms are sometimes deliberately ingested for recreational purposes although the legality of such actions varies between countries.

Finally, there is a large number of species which, when eaten, are not deadly poisonous but which can cause severe stomach upsets. Avoid the Livid Entoloma (*Entoloma sinuatum*), the Brown Roll-rim (*Paxillus involutus*), the Yellow-staining Mushroom (*Agaricus xanthodermus*), the Devil's Boletus (*Boletus satanas*), the Fairy-cake Hebeloma (*Hebeloma crustuliniforme*), the Green-spored Lepiota (*Chlorophyllum molybdites*), most *Tricholoma* species, and the peppery brittle-gills (*Russula*) and milk-caps (*Lactarius* species).

**REMEMBER**

- **If you suffer the effects of poisoning, seek medical help immediately, especially when there is a delay in the onset of symptoms.**

## Cultivated Fungi

Edible mushrooms are increasingly cultivated for Western cuisine. The Cultivated Mushroom (*Agaricus bisporus*) is produced and sold everywhere, but a number of other species are now appearing on the supermarket shelves. These include the Oyster Mushroom (*Pleurotus ostreatus*), the Shiitake Mushroom (*Lentinula edodes*), the Enoki-take (*Flammulina velutipes*), and Padi Straw Mushroom (*Volvariella volvacea*). Cultivation of the last three species has largely been developed in the Far East.

### Key to Symbols

EDIBLE

INEDIBLE

DEADLY POISONOUS

POISONOUS

EDIBILITY UNKNOWN

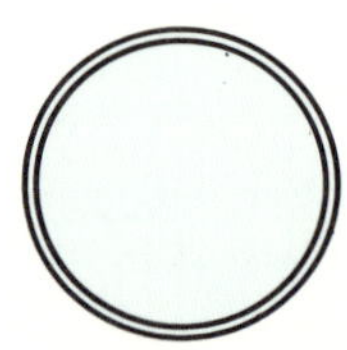
SPRING

SUMMER

AUTUMN

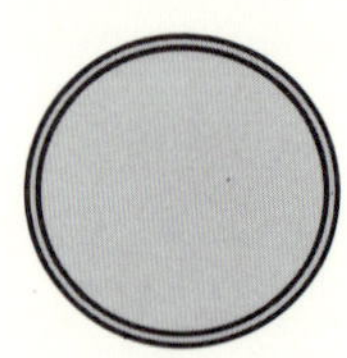
WINTER

# DIRECTORY

CHAPTER TWO

# ON THE GROUND IN WOODLANDS, OR ASSOCIATED WITH TREES

## *CANTHARELLUS CIBARIUS*

A bright orange-yellow, fleshy mushroom with thick branching ridges descending down the stem instead of thin gills.

*Cap* 1¼–6 in/3–15 cm in diameter, egg-yellow to orange, convex then flattened, with a wavy, inrolled margin, and a felty to smooth, dry surface. *Gills* decurrent, formed by thick folds and ridges. *Stem* short and tapering, 1¼–4 × 3⁄16–1 in/3–10 × 0.5–2.5 cm, solid, similarly coloured to cap. *Flesh* thick, pale yellow, with a mild smell of apricots. *Spore deposit* white. *Habitat* under beech and oak, rarely conifers. *Edibility* excellent and much sought after.

*CANTHARELLUS CIBARIUS*

**THE CHANTERELLE or THE GIROLLE**

GROUP Chanterelles

FAMILY Cantharellaceae

SEASON EDIBILITY

SIMILAR SPECIES

False Chanterelle (*Hygrophoropsis aurantiaca*, page 56) has true, thin gills and is inedible.

## *CANTHARELLUS TUBAEFORMIS*

A coniferous woodland species, often found in great numbers, but may be difficult to see in the litter; distinguished by the scaly cap, greyish and forking gills, and the yellowish stem.

*Cap* ¾–2¼ in/2–6 cm in diameter, thin, depressed with a very wavy margin, greyish brown. *Gills* reduced to narrow, decurrent, yellowish to greyish ridges and veins, which are strongly interconnected. *Stem* 1¼–2¾ × ⅛–⅜ in/3–7 × 0.3–1 cm, flattened and hollow to the base, yellowish orange to greyish, smooth. *Flesh* thin but firm, with a pleasant smell. *Spore deposit* white. *Habitat* among forest litter, occasionally on very rotten wood, more common in northerly regions. *Edibility* good; collected in great numbers throughout Scandinavia, where the fruitbodies are dried and used in stews during the winter.

*CANTHARELLUS TUBAEFORMIS*

**AUTUMN CHANTERELLE or TRUMPET CHANTERELLE**

GROUP Chanterelles

FAMILY Cantharellaceae

SEASON EDIBILITY

SIMILAR SPECIES

*C. lutescens* grows on chalky soil in pine woods in mountainous areas, and has poorly developed ridges under a blackish cap; inedible.

## *CRATERELLUS CORNUCOPIOIDES*

Dull-coloured, hollow, trumpet-shaped fruitbodies, with no ridges on the outer surface.

*Cap* ¾–3 in/2–8 cm in diameter, blackish brown, paler when dry, scaly, with a wavy margin, hollow in centre. *Outer surface* smooth, without gills, pale grey or darker, appearing powdery. *Stem* indistinct, short and hollow. *Flesh* thin, greyish, without distinctive smell. *Spore deposit* white. *Habitat* under beech or oak, forming large troops among fallen leaves. *Edibility* good, often used in stews.

*CRATERELLUS CORNUCOPIOIDES*

**HORN OF PLENTY or BLACK TRUMPET**

GROUP Chanterelles

FAMILY Cantharellaceae

SEASON EDIBILITY

SIMILAR SPECIES

The North American *C. fallax*, also edible, differs in having a strong odour, and an ochraceous spore deposit.

## *GOMPHUS CLAVATUS*

**A** thick, funnel-shaped, violaceous mushroom, having a strongly wrinkled to ridged undersurface, and growing in small clusters in coniferous forests.

*Cap* 1¼–4 in/3–10 cm in diameter, depressed with a raised wavy margin, smooth or slightly scaly, violet, eventually becoming yellowish brown. *Gills* strongly decurrent, reduced to narrow, indefinite, branching ridges, at first violet then fading. *Stem* ¾–2 × ⅜–¾ in/2–5 × 1–2 cm, short and thick, smooth, pale lilaceous brown. *Flesh* white and firm. *Spore deposit* ochre brown. *Habitat* small groups among leaf-litter; much more common in North America than in Europe where it is rare. *Edibility* good.

*GOMPHUS CLAVATUS*

**PIG'S EAR GOMPHUS**

GROUP Chanterelles

FAMILY Gomphaceae

SEASON EDIBILITY

SIMILAR SPECIES

The North American *G. floccosus* is a tall, hollow, reddish orange species, and is inedible.

## *HYGROPHORUS HYPOTHEJUS*

**A** late-season, pine-wood species with a very slimy, olive-brown cap, yellowish gills, and a slimy, pale yellow stem.

*Cap* 1¼–2¾ in/3–7 cm in diameter, convex becoming flattened or depressed at the centre, slimy, dark olive-brown, with margin paler, radially streaked. *Gills* decurrent, pale yellow, thick, widely spaced. *Stem* 2–3 × ¼–½ in/5–8 × 0.6–1.3 cm, pale yellowish to orange-yellow, slimy below a ring-like zone. *Flesh* white to pale yellow, lacking a distinctive smell. *Spore deposit* white. *Habitat* in pine woods. *Edibility* edible, but worthless.

*HYGROPHORUS HYPOTHEJUS*

**HERALD OF THE WINTER**

GROUP Mushrooms and Toadstools

FAMILY Wax Gill (Hygrophoraceae)

SEASON EDIBILITY

SIMILAR SPECIES

*H. olivaceo-albus* differs in having white gills and occurs in deciduous woods.

A comparatively rare species, it is inedible.

## *HYGROPHORUS CHRYSODON*

**A** white, slimy mushroom, with decurrent gills and yellow granules on the cap and upper stem surface. Easy to recognize, but uncommon.

*Cap* 1½–2¾ in/4–7 cm in diameter, convex with an incurved margin, shiny when dry. *Gills* white, broad, thick and waxy, well spaced. *Stem* 1¼–2¼ × 3⁄16–⅜ in/3–6 × 0.5–1 cm, white, solid, with yellow granules confined to upper region. *Flesh* thick, white, with a smell of Jerusalem artichoke. *Spore deposit* white. *Habitat* under beech and oak, sometimes conifers. *Edibility* edible.

*HYGROPHORUS CHRYSODON*

**YELLOW DOWNY HYGROPHORUS or GOLDEN-SPOTTED WAXY CAP**

GROUP Mushrooms and Toadstools

FAMILY Wax Gill (Hygrophoraceae)

SEASON EDIBILITY

SIMILAR SPECIES

*H. cossus* lacks the yellow granules; inedible.

## *LACCARIA AMETHYSTEA*

A very common, entirely deep violet-coloured species growing in troops in woodland.

*Cap* ¾–2¼ in/2–6 cm in diameter, convex then flattened, with wavy margin, slightly scurfy, deep purplish violet, powdery near the top, and with a downy, lilac coating at the base. *Gills* violet, thick, widely spaced, adnate or slightly decurrent. *Stem* 1¼–3 × 3/16–¼ in/3–8 × 0.4–0.6 cm, fibrous, tough, similarly coloured to the cap. *Flesh* thin, fibrous, pale lilac, often becoming hollow in the stem; smell not distinctive. *Spore deposit* white. *Habitat* in deciduous and coniferous woods, usually in troops. *Edibility* edible.

*LACCARIA AMETHYSTEA*

**AMETHYST DECEIVER**

**GROUP** Mushrooms and Toadstools

**FAMILY** Tricholoma (Tricholomataceae)

**SEASON** **EDIBILITY**

**SIMILAR SPECIES**

The Deceiver (*L. laccata*) and other species of *Laccaria* (see following) are of similar size and form. They differ in colour but similarly fade with age and are also edible, but tasteless.

## *LACCARIA LACCATA*

One of the most common species but very variable in appearance, recognized by the pinkish brown fruitbodies with widely spaced, "powdery" gills.

*Cap* ⅜–1½ in/1–4 cm in diameter, convex, soon becoming flattened or depressed, wavy at the margin, pinkish brown when moist, drying pale ochre or whitish, scurfy. *Gills* adnexed to slightly decurrent, pinkish brown. *Stem* 1¼–3 × 3/16–¼ in/3–8 × 0.4–0.6 cm, tall and twisted, fibrous-tough, similarly coloured to the cap. *Flesh* thin, pale, odourless. *Spore deposit* white. *Habitat* in scattered troops, among leaf-litter in woods. *Edibility* edible but tasteless.

*LACCARIA LACCATA*

**THE DECEIVER**

**GROUP** Mushrooms and Toadstools

**FAMILY** Tricholoma (Tricholomataceae)

**SEASON** **EDIBILITY**

**SIMILAR SPECIES**

*L. proxima* is more robust and often found in damper situations; edible. *L. trullisata* is commonly found growing in sand in North America; edible.

## *LEPISTA NUDA*

Large, violet to lilac fruitbodies appearing in large numbers in late autumn and winter.

*Cap* 2¾–6 in/7–15 cm in diameter, convex expanding with upturned margin, grey-violet or brownish, smooth. *Gills* sinuate, bright violet then discolouring, very crowded. *Stem* 1½–3 × ⅜–¾ in/4–8 × 1–2 cm, thick and often with a swollen base, bright violet. *Flesh* thick, pale or lilaceous, with a faint fruity smell. *Spore deposit* pale pinkish. *Habitat* among leaf-litter, sometimes forming fairy rings. *Edibility* good, often sold in European markets, although some people experience an allergic reaction.

*LEPISTA NUDA*

**WOOD BLEWIT**

**GROUP** Mushrooms and Toadstools

**FAMILY** Tricholoma (Tricholomataceae)

**SEASON** **EDIBILITY**

**SIMILAR SPECIES**

*Cortinarius alboviolaceus* develops brown gills, and has a cobweb-like veil when young; inedible.

## *LEPISTA SAEVA*

**A** large fleshy mushroom, similar to the Wood Blewit (see above), but lacking lilaceous tints in the cap and stem.

*Cap* 2–4¾ in/5–12 cm in diameter, convex, becoming flattened, dull greyish brown, smooth and dry. *Gills* sinuate, white then pale pinkish grey, broad and very crowded. *Stem* 2–4 × ⅜–¾ in/5–10 × 1–2 cm, thick, mauve to deep violaceous, fibrous-scaly, solid. *Flesh* thick and firm, white. *Spore deposit* pale pinkish. *Habitat* among leaf-litter in woodlands, but it is less commonly found than the Wood Blewit. *Edibility* excellent.

*LEPISTA SAEVA*

**BLEWIT**

**GROUP** Mushrooms and Toadstools

**FAMILY** Tricholoma (Tricholomataceae)

**SEASON** **EDIBILITY**

**SIMILAR SPECIES**

Could be confused with pale specimens of *L. nuda*.

## *LEPISTA INVERSA*

**R**ecognized by the wavy, reddish brown cap and crowded, pale yellowish gills. Formerly called *Clitocybe flaccida*.

*Cap* 1½–4 in/4–10 cm in diameter, soon depressed, becoming much paler on drying. *Gills* decurrent, thin, very crowded. *Stem* 1¼–2¼ × 3/16–⅜ in/3–6 × 0.5–1 cm, paler than cap, becoming hollow and often flattened, smooth with a hairy base. *Flesh* thin, whitish, with an earthy smell. *Spore deposit* pale pinkish cream. *Habitat* clustered among fallen leaves, sometimes forming fairy rings. *Edibility* caps good but slightly acidic to some tastes; stems too tough.

*LEPISTA INVERSA*

**TAWNY FUNNEL CAP**

**GROUP** Mushrooms and Toadstools

**FAMILY** Tricholoma (Tricholomataceae)

**SEASON** **EDIBILITY**

**SIMILAR SPECIES**

*L. gilva* is restricted to coniferous woodland, and has a pale yellowish cap; it is edible.

## *CLITOCYBE CLAVIPES*

**A** white-spored species with decurrent gills, and a conspicuously swollen stem base.

*Cap* 1¼–3 in/3–8 cm in diameter, convex, flattened or slightly depressed, greyish brown, smooth. *Gills* deeply decurrent, whitish to cream, spaced. *Stem* 1¼–2¼ × 3/16–⅜ in/3–6 × 0.5–1 cm, swollen towards the base, sometimes strongly so, greyish, spongy. *Flesh* thick and white. *Spore deposit* white. *Habitat* common under both deciduous trees and conifers. *Edibility* edible, but produces allergic reactions in some people, and should be avoided with alcohol.

*CLITOCYBE CLAVIPES*

**CLUB-FOOTED CLITOCYBE or FLAT-FOOTED CLITOCYBE**

**GROUP** Mushrooms and Toadstools

**FAMILY** Tricholoma (Tricholomataceae)

**SEASON** **EDIBILITY**

**SIMILAR SPECIES**

Unlikely to be confused with other species.

## *CLITOCYBE GEOTROPA*

**A** very large, creamy white species, having decurrent gills, and forming fairy rings in open woodland.

*Cap* 4–8 in/10–20 cm in diameter, convex to flattened but always retaining a raised centre and an incurved margin, at first yellowish brown but soon becoming whitish. *Gills* whitish, very crowded. *Stem* 2¼–8 × ⅜–1¼ in/6–20 × 1–3 cm, cylindrical, perhaps widening towards the base, solid, similarly coloured to the cap. *Flesh* white, thick and firm, with a smell of bitter almonds. *Spore deposit* white. *Habitat* in mixed woodland, preferring chalky soils, and can reappear over many years. *Edibility* edible when young but soon becoming too tough to eat.

*CLITOCYBE GEOTROPA*

**TRUMPET CLITOCYBE**

**GROUP** Mushrooms and Toadstools

**FAMILY** Tricholoma (Tricholomataceae)

**SEASON** **EDIBILITY**

**SIMILAR SPECIES**

*Leucopaxillus giganteus* is larger with a funnel-shaped cap; poisonous.

## *CLITOCYBE INFUNDIBULIFORMIS*

**A** funnel-shaped species, with ochraceous brown cap, and narrow, crowded gills running down the stem.

*Cap* 1½–2¾ in/4–7 cm in diameter, soon depressed, smooth and dry, with a very thin, wavy margin. *Gills* whitish. *Stem* 1½–3 × 3/16–⅜ in/4–8 × 0.5–1 cm, off-white, cylindrical, smooth. *Flesh* thin, pale, with a weak smell of bitter almonds. *Spore deposit* white. *Habitat* usually in deciduous woodland. *Edibility* not recommended.

*CLITOCYBE INFUNDIBULIFORMIS*

**COMMON FUNNEL CAP or FUNNEL CLITOCYBE**

**GROUP** Mushrooms and Toadstools

**FAMILY** Tricholoma (Tricholomataceae)

**SEASON** **EDIBILITY**

**SIMILAR SPECIES**

*C. squamulosa* has a scaly cap, and is found under conifers; inedible.

## *CLITOCYBE NEBULARIS*

**A** large, fleshy species which grows in large numbers among fallen leaves, and has a greyish brown cap which often appears to have a powdery bloom.

*Cap* 2¼–6 in/6–15 cm in diameter, convex to flattened, dry, smooth, often with a whitish, powdery bloom. *Gills* short decurrent, cream coloured, densely crowded. *Stem* 2¼–4¾ × ⅜–1¼ in/6–12 × 1–3 cm, often with a swollen base, pale greyish. *Flesh* thick, pale, with an unpleasant odour. *Spore deposit* pinkish cream. *Habitat* in deciduous woodland, sometimes forming fairy rings. *Edibility* may cause allergic reactions in some people.

*CLITOCYBE NEBULARIS*

**CLOUDED CLITOCYBE**

**GROUP** Mushrooms and Toadstools

**FAMILY** Tricholoma (Tricholomataceae)

**SEASON** **EDIBILITY**

**SIMILAR SPECIES**

. *geotropa* (see this page) is a larger, paler cies, with a raised centre to the cap and a ıll stem; poisonous. *Entoloma sinuatum* grows in similar situations but has yellowish gills which become pinkish; poisonous.

## CLITOCYBE ODORA

**D**istinguished by the bluish green cap, whitish gills, and the strong odour of anise.

*Cap* 1¼–2¾ in/3–7 cm in diameter, soon flattened or depressed, with an incurved, wavy margin, dingy bluish green, but sometimes blue or paling to almost white when dry, smooth. *Gills* slightly decurrent, whitish or tinged greenish, crowded. *Stem* 1¼–2¾ × 3⁄16–3⁄8 in/3–7 × 0.5–1 cm, whitish or greenish, with a woolly base. *Flesh* thin, pale, with a strong odour. *Spore deposit* white. *Habitat* prefers woodland, growing with oak; occasional. *Edibility* edible, and can be dried for use as a condiment.

*CLITOCYBE ODORA*

**BLUE-GREEN CLITOCYBE or ANISE-SCENTED CLITOCYBE**

**GROUP** Mushrooms and Toadstools

**FAMILY** Tricholoma (Tricholomataceae)

**SEASON** **EDIBILITY**

**SIMILAR SPECIES**

*C. fragrans* has a similar smell, but is a more slender, pale yellowish brown species; inedible. *C. suaveolens* is also smaller, greyish white and grows in coniferous woods; inedible.

## CLITOCYBE DICOLOR

**A** small, thin-fleshed species with decurrent gills, which is grey-brown when moist but dries out to almost white.

*Cap* ¾–2¼ in/2–6 cm in diameter, convex with a conspicuous central depression, greyish brown when water-soaked, smooth. *Gills* grey, moderately spaced. *Stem* 1¼–2¼ × ⅛–3⁄16 in/3–6 × 0.3–0.5 cm, slender, cylindrical, hollow, whitish towards the top but greyish brown below. *Flesh* thin, grey, lacking a smell. *Spore deposit* white. *Habitat* frequent in deciduous woodland. *Edibility* edible but worthless.

*CLITOCYBE DICOLOR*

**GREY-BROWN CLITOCYBE**

**GROUP** Mushrooms and Toadstools

**FAMILY** Tricholoma (Tricholomataceae)

**SEASON** **EDIBILITY**

**SIMILAR SPECIES**

*C. vibecina* has a mealy smell when crushed; inedible. *C. phyllophila* is larger, and has a lead-white to cream cap; inedible.

## LYOPHYLLUM CONNATUM

**F**orming pure white clusters on rich, peaty soil, usually at the edge of paths and in woodland clearings.

*Cap* 1¼–4 in/3–10 cm in diameter, convex with a wavy margin, pure white or pale greyish when moist, smooth. *Gills* adnate or very short decurrent, white to cream, narrow and crowded. *Stem* 1¼–3 × 3⁄16–3⁄8 in/3–8 × 0.5–1 cm, cylindrical, white to pale greyish. *Flesh* white and firm, with a distinctive alkaline smell. *Spore deposit* white. *Habitat* mixed woodland, usually on rich soil, sometimes appearing in long rows. *Edibility* edible but must first be boiled, and care must be taken to avoid confusion with the poisonous white *Clitocybe* species.

*LYOPHYLLUM CONNATUM*

**WHITE LYOPHYLLUM**

**GROUP** Mushrooms and Toadstools

**FAMILY** Tricholoma (Tricholomataceae)

**SEASON** **EDIBILITY**

**SIMILAR SPECIES**

Could be confused with poisonous white *Clitocybe* species.

## *LYOPHYLLUM FUMATOFOETENS*

**A** greyish mushroom of coniferous woodland, readily bruising blackish in all parts.

*Cap* $1\frac{1}{2}$–$3\frac{1}{2}$ in/4–9 cm in diameter, convex then flattened, sooty grey, radially finely fibrous with a shaggy margin. *Gills* adnexed, pale yellowish soon staining brownish black, crowded. *Stem* $2$–$3\frac{1}{2} \times \frac{3}{16}$–$\frac{5}{16}$ in/5–9 × 0.5–0.8 cm, cylindrical, similarly coloured to cap, longitudinally fibrous. *Flesh* white but quickly discolouring black on exposure. *Spore deposit* white. *Habitat* among leaf-litter in woods; solitary and occasional. *Edibility* inedible.

*LYOPHYLLUM FUMATOFOETENS*

**BLACKENING LYOPHYLLUM**

**GROUP** Mushrooms and Toadstools

**FAMILY** Tricholoma (Tricholomataceae)

**SEASON** **EDIBILITY**

**SIMILAR SPECIES**

Unlikely to be confused with other species.

## *TRICHOLOMA FULVUM*

**O**ne of several brown *Tricholoma* species, recognized by the slimy cap and yellowish gills and flesh.

*Cap* $2\frac{3}{4}$–$4\frac{3}{4}$ in/7–12 cm in diameter, convex to depressed, reddish brown and radially streaked, sticky. *Gills* sinuate, pale yellow soon developing reddish brown spots, crowded. *Stem* $2\frac{3}{4}$–$4\frac{3}{4} \times \frac{3}{16}$–$\frac{3}{8}$ in/7–12 × 0.5–1 cm, stocky, reddish brown, at first slimy, scaly below. *Flesh* white, yellow in stem, with a smell of rancid meal. *Spore deposit* white. *Habitat* especially coniferous and birch woods. *Edibility* inedible.

*TRICHOLOMA FULVUM*

**YELLOW-BROWN TRICHOLOMA**

**GROUP** Mushrooms and Toadstools

**FAMILY** Tricholoma (Tricholomataceae)

**SEASON** **EDIBILITY**

**SIMILAR SPECIES**

*T. albobrunneum* has a dark brown cap and white gills; inedible.

## *TRICHOLOMA VACCINUM*

**A**n uncommon species of coniferous woodland, recognized by the russet to rufous brown, scaly cap, shaggy margin, and a smell of meal.

*Cap* $1\frac{1}{2}$–$2\frac{1}{4}$ in/4–6 cm in diameter, convex with an incurved, shaggy margin, and usually with a raised centre, dry, with the surface disrupting into coarse, woolly scales. *Gills* adnexed to sinuate, whitish, discolouring pale brownish with age, crowded. *Stem* $1\frac{1}{4}$–$2 \times \frac{3}{16}$–$\frac{3}{8}$ in/3–5 × 0.8–1 cm, slender, cylindrical, fibrillose or slightly scaly, paler than the cap. *Flesh* pale, becoming hollow in the stem. *Spore deposit* white. *Habitat* in coniferous woods. *Edibility* edible but poor, with a bitter taste.

*TRICHOLOMA VACCINUM*

**RUSSET-SCALY TRICHOLOMA**

**GROUP** Mushrooms and Toadstools

**FAMILY** Tricholoma (Tricholomataceae)

**SEASON** **EDIBILITY**

**SIMILAR SPECIES**

*T. imbricatum,* also found in coniferous woods, is reddish brown but has a more convex cap which is only minutely scaly, a smooth margin, and no smell of meal; inedible.

## TRICHOLOMA USTALE

**A** deciduous wood species that has a chestnut-brown or reddish brown, smooth, sticky cap, gills that develop reddish brown spots, whitish flesh, and that lacks a distinctive smell.

*Cap* 2–3½ in/5–9 cm in diameter, convex, smooth, sticky, especially in wet weather, paler towards the margin, becoming darker with age. *Gills* adnexed to sinuate, at first white soon becoming brown spotted, crowded. *Stem* 1¼–2¼ × ⅜–⅝ in/3–6 × 1–1.5 cm, slightly thicker towards the base, whitish above, pale brown below. *Flesh* whitish, without a distinctive smell. *Spore deposit* white. *Habitat* in deciduous woods, particularly under beech. *Edibility* inedible.

*TRICHOLOMA USTALE*

**BURNT TRICHOLOMA**

GROUP Mushrooms and Toadstools

FAMILY Tricholoma (Tricholomataceae)

SEASON EDIBILITY

SIMILAR SPECIES

*T. ustaloides*, also in deciduous woods, is very similar but differs especially in having a strong smell of meal; inedible.

*T. albobrunneum* occurs in pine woods, and has a more slimy cap with white, radiating fibrils; inedible.

## TRICHOLOMA SEJUNCTUM

**An** uncommon species that has a moist, radially streaky, yellowish green cap, a white stem with a yellow tinge, and a smell of meal.

*Cap* 1½–4 in/4–10 cm in diameter, at first convex or broadly conical, finally expanded, radially streaky, yellowish green with the centre often brown or greyish brown, slimy. *Gills* sinuate, whitish, distant. *Stem* 1½–3 × ½–1 in/4–7.5 × 1–2.5 cm, whitish with yellow tints, smooth, cylindrical. *Flesh* white, firm, with a smell of meal. *Spore deposit* white. *Habitat* in deciduous woods. *Edibility* inedible.

*TRICHOLOMA SEJUNCTUM*

**SEPARATING TRICHOLOMA**

GROUP Mushrooms and Toadstools

FAMILY Tricholoma (Tricholomataceae)

SEASON EDIBILITY

SIMILAR SPECIES

*T. viridilutescens* has a more yellow cap, yellowish gills and flesh, and is a comparatively rare species; inedible.

## TRICHOLOMA SULPHUREUM

**A** uniformly sulphur yellow mushroom, with a strongly unpleasant smell reminiscent of coal-gas.

*Cap* 1½–3 in/4–8 cm in diameter, convex or with a swollen centre, smooth and dry. *Gills* sinuate, thick and well spaced. *Stem* 2¾–4 × ⅜–⅝ in/7–10 × 1–1.5 cm, cylindrical, smooth. *Flesh* firm, yellow. *Spore deposit* white. *Habitat* solitary in deciduous woods, especially on clay soils. *Edibility* distinctly unpleasant.

*TRICHOLOMA SULPHUREUM*

**SULPHUR TRICHOLOMA**

GROUP Mushrooms and Toadstools

FAMILY Tricholoma (Tricholomataceae)

SEASON EDIBILITY

SIMILAR SPECIES

*T. sejunctum* (see this page) smells of meal and has a sticky, radially streaky cap; inedible.

*T. equestre* has a sticky cap, usually with a red-brown centre, and yellowish white stem and flesh; inedible.

## *TRICHOLOMA PORTENTOSUM*

**A** large, smoke-grey *Tricholoma,* growing in large groups under pine or oak.

*Cap* 2–4¾ in/5–12 cm in diameter, conical then convex with a raised centre, and an incurved margin, grey-black, with darker radial fibres, sticky to slimy. *Gills* sinuate, white with greyish tints, narrow, fairly crowded. *Stem* 2¼–4 × ⅜–¾ in/6–10 × 1–2 cm, stout and cylindrical, solid, whitish and streaked. *Flesh* white, with a smell of meal. *Spore deposit* white. *Habitat* scattered in woodland, preferring sandy soils. *Edibility* good but avoid confusion with poisonous *Entoloma* species.

*TRICHOLOMA PORTENTOSUM*

**DINGY TRICHOLOMA or SILKY GREY TRICHOLOMA**

GROUP Mushrooms and Toadstools

FAMILY Tricholoma (Tricholomataceae)

SEASON EDIBILITY

SIMILAR SPECIES

*T. virgatum* is paler and has a bitter to peppery taste; inedible. Avoid confusion with poisonous *Entoloma* species (see *E. clypeatum,* page 36).

## *TRICHOLOMA TERREUM*

**A** coniferous-wood species, having a dark grey, fibrous-scaly cap and white gills.

*Cap* 1¼–3 in/3–8 cm in diameter, convex with a raised centre then expanding, dry, radially scaly and sometimes splitting to reveal the underlying white flesh. *Gills* sinuate, white but greyish towards the margin, broad and crowded. *Stem* 1½–2¾ × 3⁄16–⅜ in/4–7 × 0.5–1 cm, cylindrical, whitish, smooth. *Flesh* white or pale grey, with a fungoid smell. *Spore deposit* white. *Habitat* common among needle-litter. *Edibility* inedible; one should avoid all grey *Tricholoma* species.

*TRICHOLOMA TERREUM*

**GREY TRICHOLOMA**

GROUP Mushrooms and Toadstools

FAMILY Tricholoma (Tricholomataceae)

SEASON EDIBILITY

SIMILAR SPECIES

There are several closely related greyish *Tricholoma* species.

## *TRICHOLOMA SCALPTURATUM*

**A** greyish brown species, characterized by the gills and cap margin becoming yellow when bruised.

*Cap* 1½–3 in/4–8 cm in diameter, convex or flattened, pale grey with small greyish brown scales, dry. *Gills* sinuate, white becoming bright yellow, broad, spaced. *Stem* 1½–2 × 3⁄16–⅜ in/4–5 × 0.5–1 cm, short, pure white, smooth, dry. *Flesh* white, with a smell of meal. *Spore deposit* white. *Habitat* common in deciduous woodland. *Edibility* edible but not recommended.

*TRICHOLOMA SCALPTURATUM*

**YELLOW-STAINING TRICHOLOMA**

GROUP Mushrooms and Toadstools

FAMILY Tricholoma (Tricholomataceae)

SEASON EDIBILITY

SIMILAR SPECIES

The yellow staining should distinguish this mushroom from other species.

## *COLLYBIA BUTYRACEA*

**A** common species, readily recognized by the brown cap with a greasy, raised centre, white gills, and a stem which is tough, fibrous and thicker towards the base.

*Cap* $1\frac{1}{4}$–3 in/3–8 cm in diameter, convex with a raised centre, yellowish brown to dark brown, drying paler but remaining dark at the centre, moist and greasy, smooth. *Gills* adnexed to free, whitish, crowded. *Stem* $1\frac{1}{4}$–$2\frac{1}{4}$ × $\frac{3}{16}$–$\frac{3}{8}$ in/3–6 × 0.5–1 cm, thicker downwards and often swollen at the base, of similar colour to the cap, becoming hollow. *Flesh* pale, fibrous, lacking a distinctive smell. *Spore deposit* white. *Habitat* in leaf-litter in both deciduous and coniferous woods. *Edibility* edible but poor and not recommended.

*COLLYBIA BUTYRACEA*

### GREASY TOUGH SHANK

**GROUP** Mushrooms and Toadstools

**FAMILY** Tricholoma (Tricholomataceae)

**SEASON** **EDIBILITY**

**SIMILAR SPECIES**

Unlikely to be confused with other species. *C. dryophila* (see this page) grows in similar locations but has a flat, non-greasy cap and a pale, slender stem; inedible.

## *COLLYBIA CONFLUENS*

**G**reyish brown fruitbodies with tall, velvety stems, growing in dense tufts with stem bases fused.

*Cap* $\frac{3}{4}$–$1\frac{1}{2}$ in/2–4 cm in diameter, convex or flattened, thin, pinkish brown drying paler, smooth. *Gills* adnexed, whitish, narrow and very crowded. *Stem* 2–4 × $\frac{1}{8}$ in/5–10 × 0.3–0.4 cm, tall and slender, often flattened, pinkish but covered with a dense, finely hairy layer. *Flesh* thin, pale, odourless. *Spore deposit* pale cream colour. *Habitat* among dead, fallen leaves, especially in beech woods. *Edibility* worthless.

*COLLYBIA CONFLUENS*

### CLUSTERED TOUGH SHANK or TUFTED COLLYBIA

**GROUP** Mushrooms and Toadstools

**FAMILY** Tricholoma (Tricholomataceae)

**SEASON** **EDIBILITY**

**SIMILAR SPECIES**

*C. peronata* (see page 29) is more robust, yellowish brown, with spaced gills and does not grow in dense tufts; inedible.

## *COLLYBIA DRYOPHILA*

**A** small mushroom, with flattened, smooth cap, and crowded, white gills contrasting with a smooth, yellowish brown stem.

*Cap* $\frac{3}{4}$–2 in/2–5 cm in diameter, convex or flattened with a wavy margin, pale yellow to reddish brown, drying paler. *Gills* adnexed, narrow. *Stem* $1\frac{1}{4}$–$2\frac{3}{4}$ × $\frac{1}{8}$–$\frac{3}{16}$ in/3–7 × 0.3–0.5 cm, slender, shiny, hollow. *Flesh* thin, white, with a "fungussy" smell. *Spore deposit* white. *Habitat* among leaf-litter in woodlands, especially with oak, often very numerous. *Edibility* worthless and should never be eaten raw.

*COLLYBIA DRYOPHILA*

### RUSSET TOUGH SHANK or OAK-LOVING COLLYBIA

**GROUP** Mushrooms and Toadstools

**FAMILY** Tricholoma (Tricholomataceae)

**SEASON** **EDIBILITY**

**SIMILAR SPECIES**

*C. butyracea* (see this page) has a greasy centre to the cap, and has a swollen, brown stem; edible.

## *COLLYBIA MACULATA*

**A** common species which can be readily recognized by the reddish brown spotting of the white cap, stem and gills, and the very crowded, narrow gills.

*Cap* 1½–4 in/4–10 cm in diameter, convex becoming flattened and irregularly depressed, at first white soon becoming spotted with reddish brown, rarely almost completely red-brown, dry and smooth. *Gills* free to adnexed, whitish to pale cream, usually becoming spotted with reddish brown, narrow, very crowded. *Stem* 2–4 × 5/16–½ in/5–10 × 0.8–1.2 cm, tall, whitish but often spotted reddish brown, becoming hollow, tough, fibrous, and longitudinally striated. *Flesh* white, firm, lacking a distinctive smell. *Spore deposit* cream to pale pinkish. *Habitat* often in troops in both coniferous and deciduous woodland. *Edibility* inedible owing to tough texture and a rather bitter taste.

*COLLYBIA MACULATA*

**SPOTTED TOUGH SHANK**

**GROUP** Mushrooms and Toadstools

**FAMILY** Tricholoma (Tricholomataceae)

**SEASON** **EDIBILITY**

**SIMILAR SPECIES**
Unlikely to be confused with other species.

## *COLLYBIA PERONATA*

**A** thin-fleshed, yellowish brown mushroom, with a conspicuously hairy base to the stem, growing in leaf-litter.

*Cap* ¾–2 in/2–5 cm in diameter, convex then flat, yellowish to reddish brown, dry. *Gills* adnexed, pulling away from the stem as the cap expands, yellowish brown, spaced. *Stem* 2–3½ × 3/16–5/16 in/5–9 × 0.5–0.7 cm, slender, yellowish, with a woolly-hairy base, often attached to dead leaves. *Flesh* thin, tough, fibrous. *Spore deposit* white. *Habitat* common in all kinds of woodland. *Edibility* not recommended but has a distinctive peppery taste and can be dried and used as a condiment.

*COLLYBIA PERONATA*

**WOOD WOOLLY FOOT**

**GROUP** Mushrooms and Toadstools

**FAMILY** Tricholoma (Tricholomataceae)

**SEASON** **EDIBILITY**

**SIMILAR SPECIES**
Unlikely to be confused with other species.

## *MYCENA CROCATA*

**A** slender species with bright carrot-red juice staining the gills and cap, which is released when the fungus is broken.

*Cap* ¾–1¼ in/2–3 cm in diameter, bell-shaped, greyish. *Gills* adnate, sloping upwards, white with orange stains. *Stem* 2¾–4 × ⅛ in/7–10 × 0.3–0.4 cm, yellowish, smooth. *Flesh* thin and watery. *Spore deposit* white. *Habitat* in beech woods, among leaf-litter, although often attached to fallen twigs. *Edibility* worthless.

*MYCENA CROCATA*

**THE STAINER**

**GROUP** Mushrooms and Toadstools

**FAMILY** Tricholoma (Tricholomataceae)

**SEASON** **EDIBILITY**

**SIMILAR SPECIES**
*M. galopus* (see page 30) releases a white juice and does not stain orange-red, while *M. haematopus* and *M. sanguinolenta* have a blood-red juice; all inedible.

## *MYCENA EPIPTERYGIA*

**A** coniferous woodland or heathland species, with a sticky cap, and a bright yellow, slimy stem.

*Cap* ⅜–¾ in/1–2 cm in diameter, bell-shaped, brown with radiating lines. *Gills* adnate, white, with a separable jelly-like edge. *Stem* 2–4 × ⅛ in/5–10 × 0.2–0.3 cm, tall and slender, smooth, with a hairy base. *Flesh* very thin, lemon-yellow. *Spore deposit* white. *Habitat* in moss or tall grass, often numerous. *Edibility* worthless.

*MYCENA EPIPTERYGIA*

**YELLOW-STEMMED MYCENA**

**GROUP** Mushrooms and Toadstools

**FAMILY** Tricholoma (Tricholomataceae)

**SEASON** **EDIBILITY**

**SIMILAR SPECIES**

*M. viscosa* is coloured greenish, while *M. epipterygioides* grows on conifer stumps and has an unpleasant smell; both inedible.

## *MYCENA GALOPUS*

**S**mall, brittle fruitbodies, with a bell-shaped cap and a stem which releases a copious white fluid when broken.

*Cap* ⅜–¾ in/1–2 cm in diameter, pale grey-brown, darker at centre and with radial striations. *Gills* adnexed, white, moderately crowded. *Stem* 1½–2 × ⅛ in/4–5 × 0.2–0.3 cm, slender and cylindrical, hollow, brittle, greyish, smooth. *Flesh* thin, watery, with a faint smell of radish. *Spore deposit* white. *Habitat* common, found among leaf-litter in all kinds of woodland. *Edibility* worthless.

*MYCENA GALOPUS*

**MILK-DROP MYCENA**

**GROUP** Mushrooms and Toadstools

**FAMILY** Tricholoma (Tricholomataceae)

**SEASON** **EDIBILITY**

**SIMILAR SPECIES**

The variety *candida* is pure white, while *M. leucogala* (see this page) is dark blackish brown and often more robust; both inedible.

## *MYCENA LEUCOGALA*

**A** slender species which has a dark brown to blackish cap and stem, and exudes a copious white milky juice from the broken flesh.

*Cap* ⅜–¾ in/1–2 cm in diameter, bell-shaped, partly expanding with age, dark brown or blackish, with radial grooves extending from the margin to almost the centre. *Gills* adnate, pale greyish, moderately crowded. *Stem* 2–4 × ⅛ in/5–10 × 0.2–0.3 cm, similarly coloured to the cap, whitish-downy at the base, exuding a copious white milky juice when broken. *Flesh* thin, brittle, pale grey, without a distinctive smell. *Spore deposit* whitish to pale cream. *Habitat* solitary or clustered on the ground in woodland, sometimes on burnt soil. *Edibility* edible, but too small to be worthwhile.

*MYCENA LEUCOGALA*

**BLACK MILK-DROP MYCENA**

**GROUP** Mushrooms and Toadstools

**FAMILY** Tricholoma (Tricholomataceae)

**SEASON** **EDIBILITY**

**SIMILAR SPECIES**

*M. galopus* (see this page) produces a similar milky juice but differs markedly in colour; inedible.

## *MYCENA PURA*

One of the larger *Mycena* species, distinguished by the pink to lilac colour, lack of a dark edge to the gills, and a smell that reminds one of radishes.

*Cap* ¾–2¼ in/2–6 cm in diameter, convex to bell-shaped, finally expanding and becoming flattened, pink or lilac, often drying paler, smooth, striated at the margin. *Stem* 1¼–3 × 3/16–⅜ in/3–8 × 0.4–1 cm, slender but rigid, slightly thicker towards the base, similarly coloured to the cap or slightly paler, shining, whitish-downy at the base. *Flesh* thin, pinkish, with a smell of radish. *Spore deposit* white. *Habitat* among leaf-litter in deciduous woodland, especially under beech. *Edibility* poisonous.

*MYCENA PURA*

**LILAC MYCENA or PINK MYCENA**

**GROUP** Mushrooms and Toadstools

**FAMILY** Tricholoma (Tricholomataceae)

**SEASON** **EDIBILITY**

**SIMILAR SPECIES**

*M. pelianthina* is also found in beech woods but has gills with a dark violaceous edge; poisonous.

## *MELANOLEUCA MELALEUCA*

A fleshy mushroom, with a dark brown cap and brown, fibrillose stem, contrasting with white, crowded gills.

*Cap* 1¼–4 in/3–10 cm in diameter, soon flattening but retaining a knob-like centre, smoky brown drying paler, smooth. *Gills* sinuate, broad. *Stem* 1¼–3 × ⅜–¾ in/3–8 × 1–2 cm, same length as cap diameter, fibrous, often twisted, whitish, with dark brown, vertical fibrils. *Flesh* pale greyish, thick, lacking a smell. *Spore deposit* white. *Habitat* under leafy trees, although this species may also be found in meadows or even on lawns. *Edibility* edible but not recommended.

*MELANOLEUCA MELALEUCA*

**CHANGEABLE MELANOLEUCA**

**GROUP** Mushrooms and Toadstools

**FAMILY** Tricholoma (Tricholomataceae)

**SEASON** **EDIBILITY**

**SIMILAR SPECIES**

There are a number of *Melanoleuca* species, most of which are difficult to distinguish without microscopical examination. *M. cognata* (see this page) is yellowish brown, *M. grammopodia* is large and robust, and *M. brevipes* has a short stem; all inedible.

## *MELANOLEUCA COGNATA*

An uncommon species recognized by the pale stem, shiny cap and pale ochraceous to pinkish gills.

*Cap* 2¼–4 in/6–10 cm in diameter, convex becoming expanded but retaining a raised centre, yellowish brown to milky-coffee colour, dry and shiny. *Gills* adnate to sinuate, pale ochraceous to pale salmon-pink, crowded. *Stem* 2¼–5¼ × ⅜–⅝ in/6–13 × 1–1.5 cm, cylindrical, often swollen at the base, longitudinally grooved, similarly coloured to the cap but paler. *Flesh* fairly thick, pale, whitish or cream colour, with a weak smell of meal. *Spore deposit* cream. *Habitat* in open places in coniferous or, sometimes, deciduous woods. *Edibility* edible but not recommended.

*MELANOLEUCA COGNATA*

**SPRING TRICHOLOMA**

**GROUP** Mushrooms and Toadstools

**FAMILY** Tricholoma (Tricholomataceae)

**SEASON** **EDIBILITY**

**SIMILAR SPECIES**

Other *Melanoleuca* species lack the pinkish ochraceous gills.

## *AMANITA CITRINA*

**R**ecognized by the combination of a lemon-yellow or whitish cap with white scales, white gills, and stem with a ring and a very swollen, rimmed base.

*Cap* 1½–3 in/4–8 cm in diameter, convex with an incurved margin, smooth but with membranous white patches of the fragmenting veil. *Gills* free, broad, crowded. *Stem* 2¾–4¾ × ⅜–⅝ in/7–12 × 1–1.5 cm, abruptly swollen at the base, white, bearing a membranous ring near the apex and a volva which forms a prominent basal rim. *Flesh* white, thick, smelling of raw potatoes. *Spore deposit* white. *Habitat* solitary, in both pine and oak woods. *Edibility* must be avoided owing to possible confusion with the Death Cap.

*AMANITA CITRINA*

**FALSE DEATH CAP or CITRON AMANITA**

**GROUP** Mushrooms and Toadstools

**FAMILY** Amanita (Amanitaceae)

**SEASON** **EDIBILITY**

**SIMILAR SPECIES**

The Death Cap (*A. phalloides,* see page 34) has a greeny bronze, streaky cap, a faintly zoned stem, and a sac-like volva; deadly poisonous. *A. junquillea* has a ridged cap margin; poisonous.

## *AMANITA EXCELSA*

**A** tall, greyish brown *Amanita* species, with greyish scales on the cap, a ring on the stem but the volva reduced to rings of scales on the stem base.

*Cap* 3–6 in/8–15 cm in diameter, soon flattened, brown or greyish with a brown centre, sticky when moist, with scattered, powdery grey scales, and the margin lacking grooves or striations. *Gills* free, white, broad, crowded. *Stem* 3–4¾ × ⅜–¾ in/8–12 × 1–2 cm, slightly expanding downwards, white to greyish; *ring* present, white, membranous, attached to upper stem; *volva* reduced to powdery-grey scales. *Flesh* thin, white. *Spore deposit* white. *Habitat* solitary in pine or oak woods. *Edibility* avoid due to possible confusion with many *Amanita* species.

*AMANITA EXCELSA*

**TALL AMANITA**

**GROUP** Mushrooms and Toadstools

**FAMILY** Amanita (Amanitaceae)

**SEASON** **EDIBILITY**

**SIMILAR SPECIES**

*A. rubescens* bruises reddish, and is edible only after cooking, while the deadly poisonous *A. pantherina* has white scales.

## *AMANITA VAGINATA*

**A**n *Amanita* species having a smooth grey cap and a grooved margin, and a membranous white volva but no ring on the stem.

*Cap* 2–4 in/5–10 cm in diameter, at first strongly bell-shaped then expanding to flat with a knob-like centre, grey, sticky when moist, sometimes retaining white, broken patches of the veil. *Gills* free, white, broad, crowded. *Stem* 4–6 (–8) × ⅜–⅝ in/10–15 (–20) × 1–1.5 cm, tall and cylindrical, hollow, whitish, usually with grey fibrils becoming smooth; *ring* absent; *volva* present, large and forming a sac-like, membranous cup. *Flesh* thin, white. *Spore deposit* white. *Habitat* common, found in deciduous woods. *Edibility* edible but must first be carefully parboiled and water thrown away.

*AMANITA VAGINATA*

**GRISETTE**

**GROUP** Mushrooms and Toadstools

**FAMILY** Amanita (Amanitaceae)

**SEASON** **EDIBILITY**

**SIMILAR SPECIES**

There are several closely related species differing in cap colour and overall size; edibility unknown.

## *AMANITA FULVA*

Similar in stature to the Grisette but with a tawny or reddish brown cap, and often more commonly found.

*Cap* $1\frac{1}{2}$–4 in/4–10 cm in diameter, bell-shaped soon expanding to flattened or even depressed, tawny brown, smooth and shiny, with a grooved or striated margin. *Gills* free, white to creamy yellow, broad, crowded. *Stem* 4–8 × $\frac{3}{8}$–$\frac{5}{8}$ in/10–20 × 1–1.5 cm, tall, whitish, sometimes slightly scaly, hollow; *ring* absent; *volva* large, sac-like, membranous, white, with pale tawny brown marks. *Flesh* thin, white. *Spore deposit* white. *Habitat* common in woodlands, often at the side of paths. *Edibility* edible only after careful parboiling.

*AMANITA FULVA*

**TAWNY GRISETTE**

**GROUP** Mushrooms and Toadstools

**FAMILY** Amanita (Amanitaceae)

**SEASON** **EDIBILITY**

**SIMILAR SPECIES**

*A. vaginata* is very similar apart from the grey rather than brown colour; poisonous when raw (see page 32).

## *AMANITA MUSCARIA*

Perhaps the best-known wild mushroom, having a large, scarlet cap with small white scales, and a membranous ring on the stem.

*Cap* 2–9 in/5–25 cm in diameter, strongly rounded then expanding to flat and plate-like, moist and shiny, with concentric rings of small white scales which may become washed away by the rain. *Gills* free, white to pale yellow, broad and crowded. *Stem* 4–9 × $\frac{3}{8}$–1 in/10–25 × 1–2.5 cm, tall, cylindrical with a swollen base, whitish, bearing small scales of veil in rings at the base. *Flesh* thick, white, yellowish under cap cuticle. *Spore deposit* white. *Habitat* in small groups, under pine or birch. *Edibility* poisonous, containing both sweat-inducing and mild hallucinogenic poisons, which can cause delirium and coma.

*AMANITA MUSCARIA*

**FLY AGARIC**

**GROUP** Mushrooms and Toadstools

**FAMILY** Amanita (Amanitaceae)

**SEASON** **EDIBILITY**

**SIMILAR SPECIES**

The variety *regalis* is yellowish brown with yellow scales, and in North America, the variety *formosa* is orange-yellow; both poisonous.

## *AMANITA PANTHERINA*

An uncommon species, recognized by the brown cap with white scales, a flesh which does not redden on bruising, a striated cap margin, and a ring low on the stem.

*Cap* 2–4 in/5–10 cm in diameter, at first convex becoming flattened, brown or greyish brown sometimes with an olive tinge, covered with small, white, separable scales. *Gills* free, white, crowded. *Stem* $2\frac{1}{4}$–$4\frac{1}{4}$ × $\frac{5}{16}$–$\frac{5}{8}$ in/6–11 × 0.8–1.5 cm, cylindrical but slightly swollen at the base, white, bearing a ring and a volva; *ring* thin, membranous, white, not striated above, attached near the middle of the stem; *volva* forming a rim to the swollen stem base and irregular rings of scales just above it. *Flesh* thick, white, not changing colour when bruised, lacking a distinctive smell. *Spore deposit* white. *Habitat* on the ground in deciduous woods, occasionally on heathland. *Edibility* very poisonous, sometimes deadly.

*AMANITA PANTHERINA*

**THE PANTHER**

**GROUP** Mushrooms and Toadstools

**FAMILY** Amanita (Amanitaceae)

**SEASON** **EDIBILITY**

**SIMILAR SPECIES**

*A. rubescens* (see page 34) has greyish scales on the cap, and a flesh which reddens when broken or bruised; poisonous when raw. *A. excelsa* (see page 32) also has grey scales, lacks a striated margin to the cap, and has a striated ring placed higher on the stem; poisonous.

## *AMANITA PHALLOIDES*

**Deadly poisonous,** even in small amounts, always wash hands after picking. Note the streaky, olive yellow cap, white gills, ring and sac-like volva on stem.

*Cap* 2¼–4¾ in/6–12 cm in diameter, convex, yellowish green, smooth and shiny. *Gills* free, white, broad, crowded. *Stem* 2¼–4 × ⅜–¾ in/6–10 × 1–2 cm, broader below, with a swollen base, white with zig-zag pattern; *ring* membranous, hanging downwards; *volva* white, membranous. *Flesh* white, with unpleasant, sickly smell. *Spore deposit* white. *Habitat* solitary, in woodlands, especially under oak trees. *Edibility* **deadly poisonous.**

*AMANITA PHALLOIDES*

**DEATH CAP**

**GROUP** Mushrooms and Toadstools

**FAMILY** Amanita (Amanitaceae)

**SEASON** **EDIBILITY**

**SIMILAR SPECIES**

The edible true mushrooms (*Agaricus* species) can be confused but have gills which discolour from pink to blackish brown.

## *AMANITA VIROSA*

**A** rare species which is entirely white and has a shaggy, fibrous stem with a ring, and a large sac-like volva.

*Cap* 2–4 in/5–10 cm in diameter, at first convex or conical becoming expanded, smooth, white, with a silky sheen. *Gills* free, white, crowded. *Stem* 3½–4¾ × ⅜–⅝ in/9–12 × 1–1.5 cm, slender, often curved, slightly thickened downwards and swollen at the base, white, fibrous to rather shaggy, with a ring and volva; *ring* fragile, white, soon frayed and torn; *volva* sac-like, membranous, white or with a greyish tinge. *Flesh* white, unchanging when bruised, with a sweetish, nauseating smell. *Spore deposit* white. *Habitat* solitary on the ground in deciduous woods, often more common in northerly areas. *Edibility* **deadly poisonous.**

*AMANITA VIROSA*

**DESTROYING ANGEL**

**GROUP** Mushrooms and Toadstools

**FAMILY** Amanita (Amanitaceae)

**SEASON** **EDIBILITY**

**SIMILAR SPECIES**

*A. citrina* var. *alba* has greyish ochre scales on the cap, a smooth stem which lacks a sac-like volva, and a smell of potatoes; poisonous. *A. ovoidea* is more robust, with well-spaced gills, and a tough, pale ochraceous volva; poisonous. Other white *Amanita* species have markedly scaly caps and lack a sac-like volva.

## *AMANITA RUBESCENS*

**A** very common, stocky *Amanita* species having a reddish brown cap with small greyish scales, and a flesh which turns pinkish on exposure.

*Cap* 2–6 in/5–15 cm in diameter, strongly convex then finally flat, grey to reddish brown, paler at the margin which is not striated, bearing small, greyish scales of the veil. *Gills* free, white but bruising reddish brown, broad, crowded. *Stem* 2–4 × 3⁄16–¾ in/5–10 × 0.5–2 cm, stocky, expanding downwards, with a swollen base, pale above, reddish brown below; somewhat scaly below the ring, smooth above; *ring* present, membranous, large, on upper stem; *volva* absent at maturity. *Flesh* thick, firm, white discolouring pinky brown on exposure, especially around insect borings. *Spore deposit* white. *Habitat* common in woods. *Edibility* edible only after cooking but avoid confusion with *A. pantherina*.

*AMANITA RUBESCENS*

**THE BLUSHER**

**GROUP** Mushrooms and Toadstools

**FAMILY** Amanita (Amanitaceae)

**SEASON** **EDIBILITY**

**SIMILAR SPECIES**

*A. pantherina* (see page 33) is less common, has white velar scales, and a rim-like volva; deadly poisonous. In North America, *A. flavorubescens* has a yellowish cap and veil; poisonous.

## *LIMACELLA GLIODERMA*

**A** white-spored species with a slimy, brown cap, white gills, and a small ring around the stem.

*Cap* 1¼–3 in/3–8 cm in diameter, convex to flattened with a raised centre, dark reddish brown, fading to pinkish, slimy. *Gills* free or adnexed, white to pinkish, broad and crowded. *Stem* 2–4 × 3/16–⅜ in/5–10 × 0.5–1 cm, pale brown, silky, dry, but with remnants of a sticky veil; *ring* present, fibrous but ephemeral. *Flesh* thin, white, with a strong smell of meal. *Spore deposit* white. *Habitat* associated with coniferous trees and occasionally with birch. *Edibility* unknown.

*LIMACELLA GLIODERMA*

**SLIMY VEIL LIMACELLA**

**GROUP** Mushrooms and Toadstools

**FAMILY** Amanita (Amanitaceae)

**SEASON** **EDIBILITY** ?

**SIMILAR SPECIES**

*L. lenticularis* is paler and has a conspicuous ring.

## *MARASMIUS WYNNEI*

**S**mall, tough mushroom, growing in dense clusters among beech litter, where it forms an extensive white mat of mycelium binding together the leaves.

*Cap* ⅜–2 in/1–5 cm in diameter, strongly convex then slowly expanding, white to pale greyish violet, with a translucent margin. *Gills* adnexed, white or pale clay coloured, narrow, spaced. *Stem* 1½–3 × ⅛ in/4–8 × 0.3–0.4 cm, slender, tough, smooth, white above, reddish brown below. *Flesh* thin, tough, white. *Spore deposit* white. *Habitat* found among leaf-litter in beech woods. *Edibility* worthless.

*MARASMIUS WYNNEI*

**WYNNE'S MARASMIUS**

**GROUP** Mushrooms and Toadstools

**FAMILY** Tricholoma (Tricholomataceae)

**SEASON** **EDIBILITY**

**SIMILAR SPECIES**

*M. cohaerens* also grows among beech leaves, but has a brown, velvety cap; inedible.

## *MARASMIUS ALLIACEUS*

**A** tall, slender and tough species of beech woods, with a blackish, velvety stem and a strong smell of garlic.

*Cap* ¾–1½ in/2–4 cm in diameter, bell-shaped and only slowly expanding, white to greyish brown, with a striated margin, dry and smooth. *Gills* adnexed, whitish, narrow, spaced. *Stem* 4–8 × ⅛ in/10–20 × 0.3–0.4 cm, tall and slender, cylindrical, hollow, black, finely ridged, velvety with very fine, white, short hairs, and a rooting base. *Flesh* thin, watery, with a very strong, garlic-like smell which can be detected at some distance. *Spore deposit* white. *Habitat* solitary among leaf-litter. *Edibility* worthless.

*MARASMIUS ALLIACEUS*

**GARLIC MUSHROOM**

**GROUP** Mushrooms and Toadstools

**FAMILY** Tricholoma (Tricholomataceae)

**SEASON** **EDIBILITY**

**SIMILAR SPECIES**

*M. scorodonius* is also known as the Garlic Mushroom, owing to its odour, but the cap is yellowish brown and the stem is pale and smooth; inedible.

## ENTOLOMA CLYPEATUM

A fleshy, stocky, spring-time mushroom with pink gills, growing in groups under rosaceous shrubs.

*Cap* 1½–3 in/4–8 cm in diameter, convex with a raised centre, greyish brown, sometimes dark, smooth and radially streaky. *Gills* sinuate, pale at first becoming pinkish, crowded. *Stem* 1½–3 × 3/16–¾ in/4–8 × 0.5–2 cm, short cylindrical, whitish tinged with brown, solid, with a fibrous surface. *Flesh* thick, white, with a smell of meal. *Spore deposit* pink. *Habitat* on the ground, under rosaceous trees, sometimes forming fairy rings. *Edibility* poisonous and can cause considerable stomach upsets, especially if eaten raw.

*ENTOLOMA CLYPEATUM*

**ROMAN SHIELD ENTOLOMA**

**GROUP** Mushrooms and Toadstools

**FAMILY** Entoloma (Entolomataceae)

**SEASON** **EDIBILITY**

**SIMILAR SPECIES**

*E. sinuatum* has a paler cap and yellowish gills, and *E. aprile* has a cap with a raised centre; both poisonous.

## CLITOPILUS PRUNULUS

A whitish species, with decurrent gills, and a strong smell and taste of bread dough.

*Cap* 1¼–4 in/3–10 cm in diameter, convex to depressed with a wavy margin, white or with a greyish tint, dry and smooth. *Gills* deeply decurrent, white then pale pink, widely spaced. *Stem* 1¼–2 × ⅜–⅝ in/3–5 × 1–1.5 cm, short, sometimes excentric, white, smooth, solid. *Flesh* thick, white, with a very strong odour. *Spore deposit* salmon-pink. *Habitat* forms small groups on ground in open woodland and grassy glades. *Edibility* excellent.

*CLITOPILUS PRUNULUS*

**THE MILLER or SWEETBREAD MUSHROOM**

**GROUP** Mushrooms and Toadstools

**FAMILY** Entoloma (Entolomataceae)

**SEASON** **EDIBILITY**

**SIMILAR SPECIES**

The poisonous *Clitocybe dealbata* grows in grassland, but has white, crowded gills, and lacks the mealy smell.

## LEPIOTA CRISTATA

The small white caps have concentric rings of dark reddish brown scales, the gills are white, and there is a ring on the stem.

*Cap* ¾–1½ in/2–4 cm in diameter, convex to almost flat, often with a raised centre, white except for the dark centre and rings of small scales. *Gills* free, narrow, and very crowded. *Stem* ¾–2¾ × ⅛–3/16 in/2–7 × 0.2–0.5 cm, slender, hollow, white and smooth, with a small, fragile ring attached to the upper region. *Flesh* thin, white, with an unpleasant, rubbery smell. *Spore deposit* white. *Habitat* grows in small troops among leaf-litter or short grass, often at the edge of woods. *Edibility* inedible and possibly poisonous.

*LEPIOTA CRISTATA*

**STINKING PARASOL and MALODOROUS LEPIOTA**

**GROUP** Mushrooms and Toadstools

**FAMILY** Mushroom and Lepiota (Agaricaceae)

**SEASON** **EDIBILITY**

**SIMILAR SPECIES**

*L. clypeolaria* is larger, with pale yellowish brown scales on both cap and stem, while *L. castanea* (see page 37) has chestnut-brown scales, and a different smell; both poisonous.

## *LEPIOTA CASTANEA*

**A** small *Lepiota* species, with a brown, scaly cap and stem, and occurring in deciduous woodland.

*Cap* ¾–1½ in/2–4 cm in diameter, convex with a raised centre, chestnut-brown, dry, with small, granular but indistinct scales. *Gills* free, cream-coloured, staining red-brown with age, crowded. *Stem* ¾–1½ × ⅛ in/2–4 × 0.2–0.4 cm, cylindrical with slightly swollen base, finely scaly below an inconspicuous ring-zone near the top. *Flesh* thin, pale yellowish, with a faintly fruity smell. *Spore deposit* white. *Habitat* found on rich soil, growing under deciduous trees. *Edibility* poisonous.

*LEPIOTA CASTANEA*

**CHESTNUT PARASOL**

**GROUP** Mushrooms and Toadstools

**FAMILY** Mushroom and Lepiota (Agaricaceae)

SEASON EDIBILITY

SIMILAR SPECIES

*L. fulvella* has a yellowish brown cap which is less scaly, while *L. ignicolor* is orange, with yellow gills; both poisonous.

## *LEPIOTA SUBINCARNATA*

**O**ne of several brown *Lepiota* species, which are distinguished on the basis of colour and scaliness of the cap.

*Cap* ⅜–1¼ in/1–3 cm in diameter, convex becoming flattened, rose-pink, with minute, crowded, reddish brown scales at the centre which become sparser towards the margin. *Gills* free, white, narrow, crowded. *Stem* ¾–1½ × ⅛ in/2–4 × 0.2–0.3 cm, cylindrical, hollow, white then discolouring reddish brown, with tiny, white, cottony scales, and an indefinite ring-zone on the upper region. *Flesh* thin, white with a pinkish tint, and a weak, fruity smell. *Spore deposit* white. *Habitat* isolated or in small groups among leaf-litter in woodland. *Edibility* poisonous.

*LEPIOTA SUBINCARNATA*

**PINKISH PARASOL**

**GROUP** Mushrooms and Toadstools

**FAMILY** Mushroom and Lepiota (Agaricaceae)

SEASON EDIBILITY

SIMILAR SPECIES

*L. fulvella* has an orange-brown cap which does not break up into distinct scales, while *L. castanea* (see this page) is chestnut-brown with small, granular scales; both poisonous.

## *MELANOPHYLLUM ECHINATUM*

**A** small greyish brown mushroom with a granular cap and dark reddish brown gills.

*Cap* ⅜–1¼ in/1–3 cm in diameter, cone-shaped to almost flat, dark greyish brown, covered with loose, granular to spiny scales which eventually become loosened and powdery, with a fringed margin. *Gills* free, reddish becoming progressively darker, broad, moderately crowded. *Stem* ¾–2¼ × ⅛ in/2–6 × 0.2–0.3 cm, short, with a swollen base, hollow and fragile, reddish brown, finely powdery; *ring* reduced to a thin zone on the upper part of the stem. *Flesh* thin, dark. *Spore deposit* reddish brown. *Habitat* occasionally found solitary among leaf-litter in deciduous woods, also found on compost heaps and in greenhouses. *Edibility* inedible.

*MELANOPHYLLUM ECHINATUM*

**RED-GILLED AGARIC**

**GROUP** Mushrooms and Toadstools

**FAMILY** Mushroom and Lepiota (Agaricaceae)

SEASON EDIBILITY

SIMILAR SPECIES

Unlikely to be confused with other species.

## CYSTODERMA AMIANTHINUM

**A** small, slender parasol mushroom, with a granular-scaly, ochre-yellow cap and stem, and white gills.

*Cap* ¾–2 in/2–5 cm in diameter, conical to convex with a raised centre, dry, granular-scaly, with a shaggy margin. *Gills* adnate, white, narrow, crowded. *Stem* 1½–3 × ⅛ in/4–8 × 0.3–0.4 cm, tall and slender, hollow, granular-scaly below the ring, smooth above; *ring* small, membranous, granular on the underside, pointing upwards. *Flesh* thin, pale, with an unpleasant mouldy smell. *Spore deposit* white. *Habitat* among conifer litter, and also found in mossy heathland. *Edibility* inedible.

*CYSTODERMA AMIANTHINUM*

### SAFFRON PARASOL or PUNGENT CYSTODERMA

GROUP Mushrooms and Toadstools

FAMILY Mushroom and Lepiota (Agaricaceae)

SEASON EDIBILITY

SIMILAR SPECIES

In North America, the variety *rugosoreticulatum* has a strongly wrinkled cap. Less common related species include *C. granulosum* which is darker reddish brown, and *C. carcharias* which is pinkish grey, with a more prominent ring on the stem. All inedible.

## MACROLEPIOTA PROCERA

**A** large mushroom, with a woolly, scaly cap and a tall, scaly stem with a large, movable ring; young unexpanded fruitbodies resembling drum-sticks.

*Cap* 3–8 in/8–20 cm in diameter, convex to flat with a central raised darker area, covered with shaggy, grey-brown scales. *Gills* free, white, broad, very crowded. *Stem* 4–15¾ × ⅜–⅝ in/10–40 × 1–1.5 cm, cylindrical except for a swollen base, covered by a "snake-skin pattern" of fibrous scales, bearing a large, movable, double-edged, membranous ring in upper region. *Flesh* thick, white, not reddening. *Spore deposit* white. *Habitat* in groups, often forming fairy rings, in grassy woodland glades, sometimes also in meadows. *Edibility* one of the best edible species.

*MACROLEPIOTA PROCERA*

### PARASOL MUSHROOM

GROUP Mushrooms and Toadstools

FAMILY Mushroom and Lepiota (Agaricaceae)

SEASON EDIBILITY

SIMILAR SPECIES

*M. gracilenta* has much smaller, indistinct scales on the cap; edible. *M. rhacodes* (see this page) has reddening flesh and lacks the snake-like pattern on the stem; inedible.

## MACROLEPIOTA RHACODES

**A** fleshy mushroom with a coarsely scaly cap, and a stem and flesh which bruises reddish.

*Cap* 2¾–4¾ in/7–12 cm in diameter, convex to flat, pale clay brown, with surface breaking up into large, fibrous scales, except at the centre. *Gills* free, white, broad, very crowded. *Stem* 4–6 × ⅜–1 in/10–15 × 1–2.5 cm, thick and expanding towards the base, smooth, off-white but soon bruising, bearing a large, double-edged, movable ring. *Flesh* thick, white discolouring saffron-red when broken. *Spore deposit* white. *Habitat* in small groups, usually on disturbed soil, at edge of woods or in gardens. *Edibility* edible but causes an allergic reaction with some persons.

*MACROLEPIOTA RHACODES*

### SHAGGY PARASOL

GROUP Mushrooms and Toadstools

FAMILY Mushroom and Lepiota (Agaricaceae)

SEASON EDIBILITY

SIMILAR SPECIES

In North America, *Chlorophyllum molybdites* is similar but has greenish gills and spore deposit, and the flesh does not discolour reddish; poisonous.

## *AGARICUS AUGUSTUS*

One of the largest of the true mushrooms, with a yellowish brown, scaly cap, bruising deep yellow when rubbed.

*Cap* 4–8 in/10–20 cm in diameter, almost spherical then expanding, retaining flattened top, covered with small, tawny brown, fibrous scales in concentric rings on a white to yellowish background, bruising deep yellow. *Gills* free, pinkish to blackish brown, crowded. *Stem* 4–8 × ¾–1¼ in/10–20 × 2–3 cm, thicker towards the base, hollow, white, bruising yellow, smooth above the ring, soft-scaly below the ring; *ring* large, membranous, white, attached towards the top of the stem. *Flesh* white, with a pleasant smell of bitter almonds. *Spore deposit* brownish black. *Habitat* on forest floor, often under spruce, or in parks under deciduous trees. *Edibility* excellent.

*AGARICUS AUGUSTUS*

**THE PRINCE**

**GROUP** Mushrooms and Toadstools

**FAMILY** Mushroom and Lepiota (Agaricaceae)

**SEASON** **EDIBILITY**

**SIMILAR SPECIES**

Unlikely to be confused with other species, but care should be taken in identifying yellow-staining species.

## *AGARICUS MACROSPORUS*

A large, fleshy, white mushroom, covered with small, loose, flaky scales, and growing in grassy glades.

*Cap* 4–6 in/10–15 cm in diameter, strongly convex then expanding and flattened, white, discolouring pale yellow in dry weather, and often cracking at the centre, shining, bearing flaky, white scales, and bruising pale yellow. *Gills* free, pink to brownish black, narrow, crowded. *Stem* 2–4 × ¾–1¼ in/5–10 × 2–3.5 cm, short and thick, often with a rooting base, solid or nearly so, white, covered with loose, white scales; *ring* large, pendulous, smooth above, scaly below. *Flesh* thick, firm, slowly developing a smell of mouldy straw. *Spore deposit* brownish black. *Habitat* growing in groups, at times forming rings, in grassy areas. *Edibility* edible.

*AGARICUS MACROSPORUS*

**LARGE-SPORED MUSHROOM**

**GROUP** Mushrooms and Toadstools

**FAMILY** Mushroom and Lepiota (Agaricaceae)

**SEASON** **EDIBILITY**

**SIMILAR SPECIES**

*A. excellens* has a tall stem, and grows in spruce forests; edible.

## *AGARICUS SILVATICUS*

A true mushroom found in coniferous woods, and recognized by the dark brown scaly cap, pink to blackish brown gills, and a flesh that darkens to blood-red when broken open.

*Cap* 2¼–4 in/6–10 cm in diameter, strongly convex then flattening, with large, fibrous scales and a thin margin. *Gills* free, thin, broad and crowded, becoming darker with age. *Stem* 2¾–4 × ⅜–¾ in/7–10 × 1–2 cm, cylindrical, slightly swollen at the base, pale brown. *Flesh* firm and white, immediately discolouring. *Spore deposit* blackish brown. *Habitat* among conifer needles on woodland floor. *Edibility* edible but not recommended.

*AGARICUS SILVATICUS*

**SCALY WOOD MUSHROOM**

**GROUP** Mushrooms and Toadstools

**FAMILY** Mushroom and Lepiota (Agaricaceae)

**SEASON** **EDIBILITY**

**SIMILAR SPECIES**

*A. haemorrhoidarius* is larger and not confined to conifer woods, while *A. silvicola* is creamy white; both edible.

## COPRINUS DISSEMINATUS

**A** tiny, bell-shaped ink-cap, growing in enormous numbers over the substratum.

*Cap* $\frac{3}{16}$–$\frac{3}{8}$ in/0.5–1 cm in diameter, pale yellowish brown slowly becoming greyish, finely grooved, bearing minute hairs which are visible with a hand-lens. *Gills* adnate, at first white then progressively grey to black from the edge, broad, moderately spaced. *Stem* $\frac{3}{8}$–$1\frac{1}{4}$ × $\frac{1}{16}$ in/1–3 × 0.1–0.2 cm, slender and brittle, fragile, hollow, white, smooth. *Flesh* thin and watery. *Spore deposit* dark brown. *Habitat* commonly found in great numbers over fallen, woodland debris, and can also be found in grassy areas. *Edibility* worthless.

*COPRINUS DISSEMINATUS*

**TROOPING CRUMBLE CAPS or NON-INKY COPRINUS**

GROUP Mushrooms and Toadstools

FAMILY Ink-cap (Coprinaceae)

SEASON EDIBILITY

SIMILAR SPECIES

Unlikely to be confused with other species.

## COPRINUS MICACEUS

**F**orming large clusters, the cap dissolving into a black inky fluid; the caps are tawny brown, distinctly grooved and glistening, and the stems are white.

*Cap* $\frac{3}{4}$–$1\frac{1}{2}$ in/2–4 cm in diameter, bell-shaped, with a glistening surface when young and fresh. *Gills* almost free, at first pale then blackening and liquefying from the edge, crowded. *Stem* 2–3 × $\frac{1}{8}$–$\frac{3}{16}$ in/5–8 × 0.3–0.5 cm, cylindrical, white, smooth, hollow and fragile. *Flesh* thin, watery. *Spore deposit* blackish brown. *Habitat* large clusters on or near stumps or buried roots. *Edibility* worthless.

*COPRINUS MICACEUS*

**GLISTENING INK-CAP or MICA CAP**

GROUP Mushrooms and Toadstools

FAMILY Ink-cap (Coprinaceae)

SEASON EDIBILITY

SIMILAR SPECIES

*C. disseminatus* (see this page) is much smaller and does not liquefy, and *C. domesticus* has white scales on cap apex, and grows on wood in small clusters; both inedible.

## COPRINUS PICACEUS

**A** solitary, robust ink-cap of beech woods, recognized by the conical black caps with large white patches of the veil.

*Cap* $1\frac{1}{2}$–$2\frac{3}{4}$ in/4–7 cm in diameter, conical and only partly expanding, covered at first by a felt-like, white veil which cracks on expansion to leave large white patches on a dark brown to blackish background. *Gills* free, white becoming black from the edge, and liquefying into an inky fluid. *Stem* $4\frac{3}{4}$–8 × $\frac{3}{16}$–$\frac{3}{8}$ in/12–20 × 0.5–1 cm, tall, cylindrical, fragile, hollow, white and smooth. *Flesh* thin, watery, with a mild but unpleasant smell. *Spore deposit* black. *Habitat* beech woods, usually on chalky soil. *Edibility* inedible.

*COPRINUS PICACEUS*

**MAGPIE INK-CAP**

GROUP Mushrooms and Toadstools

FAMILY Ink-cap (Coprinaceae)

SEASON EDIBILITY

SIMILAR SPECIES

*C. comatus* is tufted and the cap has recurved scales on a whitish background; edible.

## *PSATHYRELLA CANDOLLEANA*

**A** brittle mushroom, which grows in small groups, and has a pale yellowish cream cap with delicate, white remains of veil attached to margin, dark gills and a white stem.

*Cap* ¾–3 in/2–8 cm in diameter, convex, expanding to almost flat, smooth, fading to almost white on drying. *Gills* adnexed, pinkish grey becoming blackish brown, moderately crowded. *Stem* 1½–3 × ⅛–5⁄16 in/4–8 × 0.3–0.8 cm, cylindrical, fragile, hollow, smooth. *Flesh* thin, watery, white, brittle. *Spore deposit* purplish brown. *Habitat* usually growing in small groups in grassy areas, often near to tree-stumps. *Edibility* edible but worthless.

*PSATHYRELLA CANDOLLEANA*

**FRINGED CRUMBLE CAP or COMMON PSATHYRELLA**

GROUP Mushrooms and Toadstools

FAMILY Ink-cap (Coprinaceae)

SEASON EDIBILITY

SIMILAR SPECIES

*P. gracilis* grows in similar situations but is taller, with a more slender stem and a bell-shaped cap; inedible.

## *PSATHYRELLA SQUAMOSA*

**A**lso known as *P. artimesiae,* this fragile species is recognized by the cottony scales on the cap and stem.

*Cap* ⅜–1½ in/1–4 cm in diameter, bell-shaped to convex, reddish brown drying paler but covered with a loose veil of cottony, white fibres, especially towards the margin. *Gills* adnate, greyish brown with a white edge, fairly crowded. *Stem* ¾–2¾ × ⅛ in/2–7 × 0.2–0.4 cm, cylindrical, fragile, white, hollow, covered with a whitish down. *Flesh* thin, dark grey-brown. *Spore deposit* purplish black. *Habitat* growing on the ground in coniferous and deciduous woodlands. *Edibility* inedible.

*PSATHYRELLA SQUAMOSA*

**SCALY CRUMBLE CAP**

GROUP Mushrooms and Toadstools

FAMILY Ink-cap (Coprinaceae)

SEASON EDIBILITY

SIMILAR SPECIES

*P. pennata* grows on burnt ground but otherwise appears very similar unless viewed with a microscope; inedible.

## *STROPHARIA AERUGINOSA*

**A** common and distinctive toadstool having, when young and fresh, a slimy, deep blue-green cap with white scales at the margin, and the stem of which is white and scaly below a spreading ring.

*Cap* 1¼–3 in/3–8 cm in diameter, conical or convex at first, becoming flattened, often with a raised centre, very slimy, green or blue-green when young, with small white scales especially near the margin, the colour fading to pale yellowish with age. *Gills* adnate, at first white then purplish clay-brown, with white edges. *Stem* 1½–4 × ⅛–⅜ in/4–10 × 0.4–1 cm, cylindrical, slightly thickened at the base, hollow, white or pale green, slimy, bearing a membranous ring below which it is covered in white, cottony scales. *Flesh* whitish, usually with a green tint in the stem; lacking a distinctive smell. *Spore deposit* purplish brown. *Habitat* in mixed woodland. *Edibility* poisonous.

*STROPHARIA AERUGINOSA*

**VERDIGRIS AGARIC or GREEN STROPHARIA**

GROUP Mushrooms and Toadstools

FAMILY Stropharia (Strophariaceae)

SEASON EDIBILITY

SIMILAR SPECIES

*S. caerulea,* in urban habitats in Europe, has a scanty veil, while *S. albocyanea* is smaller, more slender and paler; both poisonous.

## *PSILOCYBE SQUAMOSA*

**A** tall, slender, yellowish orange toadstool, having a scaly cap, and growing among beech litter.

*Cap* ¾–2¼ in/2–6 cm in diameter, strongly convex then flattened, ochraceous yellow to almost orange at the centre, sticky, bearing rings of small, pointed scales. *Gills* anate to almost decurrent, grey to violaceous black, crowded. *Stem* 2¼–6 × 3/16–5/16 in/6–15 × 0.5–0.8 cm, tall, cylindrical, hollow, at first whitish then reddish brown, with small, fibrous scales. *Flesh* thin, pale. *Spore deposit* sooty black. *Habitat* among litter in beech woods. *Edibility* inedible.

*PSILOCYBE SQUAMOSA*

**SCALY PSILOCYBE**

GROUP Mushrooms and Toadstools

FAMILY Strophario (Strophariaceae)

SEASON — EDIBILITY

SIMILAR SPECIES

*P. thrausta* is more brightly coloured, and the cap has a raised centre; inedible.

## *HEBELOMA CRUSTULINIFORME*

**A** common, clay-brown species which has a strong smell of radish and exudes droplets from the gills in damp weather.

*Cap* 1½–3 in/4–8 cm in diameter, convex then expanding, with the margin remaining inrolled, buff to clay-brown, paler towards the margin, moist, smooth. *Gills* sinuate, pale clay-brown, crowded, exuding watery droplets in damp weather, dark brown spotted when dry. *Stem* 1½–2¾ × ⅜–¾ in/4–7 × 1–2 cm, cylindrical, thickened at the base, whitish, powdery especially towards the apex. *Flesh* thick, whitish, with a smell of radish. *Spore deposit* rusty brown. *Habitat* in deciduous woodland. *Edibility* poisonous, with a bitter taste.

*HEBELOMA CRUSTULINIFORME*

**FAIRY CAKE HEBELOMA or POISON PIE**

GROUP Mushrooms and Toadstools

FAMILY Cortinarius (Cortinariaceae)

SEASON — EDIBILITY

SIMILAR SPECIES

*H. longicaudum* is larger, with a long stem and lacks the radish smell; poisonous.

## *HEBELOMA MESOPHAEUM*

**O**ne of the few *Hebeloma* species with a ring-zone on the upper stipe; a small fruit-body most often associated with conifer woods.

*Cap* ¾–1½ in/2–4 cm in diameter, convex then flattened, clay-brown to chestnut-brown, paler towards the margin, sticky, smooth. *Gills* sinuate, pinkish brown to cinnamon-brown, crowded. *Stem* 1½–2¾ × ⅛ in/4–7 × 0.3–0.4 cm, cylindrical, white with brownish tints, bearing in the upper part a ring-like zone left by the cobweb-like, cortinoid veil. *Flesh* white, but characteristically brown at stem base; with a smell of radishes. *Spore deposit* clay-brown. *Habitat* mostly under conifer sometimes birch, often on burnt ground. *Edibility* poisonous, very bitter.

*HEBELOMA MESOPHAEUM*

**PINE HEBELOMA**

GROUP Mushrooms and Toadstools

FAMILY Cortinarius (Cortinariaceae)

SEASON — EDIBILITY

SIMILAR SPECIES

Unlikely to be confused with other species.

## *HEBELOMA SACCHARIOLENS*

**A** moderate-sized species that grows in deciduous woods and is characterized by a strong, sweetish smell.

*Cap* ¾–2 in/2–5 cm in diameter, convex, pale becoming more brownish with age, smooth, sticky. *Gills* sinuate, cinnamon-brown, crowded. *Stem* 1½–2 × ⅛ in/4–5 × 0.3–0.4 cm, cylindrical, whitish with tawny-brown base, smooth. *Flesh* white, firm, with a smell described as recalling orange-flowers, burnt sugar or cheap scented soap. *Spore deposit* clay-brown. *Habitat* in small groups, under deciduous trees, preferring sandy soil. *Edibility* poisonous, causing stomach upsets.

*HEBELOMA SACCHARIOLENS*

**SCENTED HEBELOMA**

GROUP Mushrooms and Toadstools

FAMILY Cortinarius (Cortinariaceae)

SEASON EDIBILITY

SIMILAR SPECIES

Unlikely to be confused with other species due to its distinctive smell.

## *HEBELOMA SINAPIZANS*

**T**his is the largest of the *Hebeloma* species, with a distinctive smell of raw potatoes.

*Cap* 2¾–6 in/7–15 cm in diameter, fleshy, convex, pale pinkish brown, smooth, slightly sticky. *Gills* sinuate, cinnamon-brown, broad, crowded. *Stem* 2¾–4¾ × ⅜–⅝ in/7–12 × 1–1.5 cm, cylindrical, hollow, white, with the surface breaking up into small curved scales. *Flesh* white, thick; in section the flesh of the cap forms a V-shaped projection into the cavity of the stem. *Spore deposit* clay-brown. *Habitat* among leaf-litter in deciduous woods. *Edibility* poisonous, causing stomach upsets.

*HEBELOMA SINAPIZANS*

**CLAYEY HEBELOMA**

GROUP Mushrooms and Toadstools

FAMILY Cortinarius (Cortinariaceae)

SEASON EDIBILITY

SIMILAR SPECIES

Unlikely to be confused with other species.

## *HEBELOMA RADICOSUM*

**A**n unusual *Hebeloma* species distinguished by having a membranous ring on the stem and a rooting base.

*Cap* 2¼–3½ in/6–9 cm in diameter, convex, pale brown, sticky, with small, inconspicuous white scales. *Gills* sinuate, cinnamon-brown, crowded. *Stem* 2–3 × ⅜–⅝ in/5–8 × 1–1.5 cm, cylindrical, with a deep rooting base, whitish, covered with indefinite fibrous scales and bearing a ring. *Flesh* white, firm. *Spore deposit* clay-brown. *Habitat* occasional, tends to grow around old stumps in beech and oak woods. *Edibility* poisonous.

*HEBELOMA RADICOSUM*

**ROOTING PHOLIOTA**

GROUP Mushrooms and Toadstools

FAMILY Cortinarius (Cortinariaceae)

SEASON EDIBILITY

SIMILAR SPECIES

Unlikely to be confused with other species.

## *INOCYBE FASTIGIATA*

**A** very common, large *Inocybe* species, with a conical, yellowish brown cap, and a smell of mouldy bread.

*Cap* 2–2¾ in/5–7 cm in diameter, strongly conical, often splitting at the margin, straw-yellow to yellowish ochre, with radial fibrils which separate to reveal a paler background. *Gills* adnate, clay-brown, often with an olive tint, narrow, crowded. *Stem* 1½–4 × 3⁄16–5⁄16 in/4–10 × 0.5–0.8 cm, tall, cylindrical, white to ochre-brown, fibrous, hollow. *Flesh* pale, firm, with a distinctive smell. *Spore deposit* pale clay-brown. *Habitat* always under trees, often beech. *Edibility* **causes muscarine poisoning and must be avoided.**

*CORTINARIUS AMOENOLENS*

**PEAKED INOCYBE or STRAW-COLORED FIBER HEAD**

GROUP Mushrooms and Toadstools

FAMILY Cortinarius (Cortinariaceae)

SEASON EDIBILITY

SIMILAR SPECIES

*I. maculata* has darker, reddish brown fibres on the cap and white veil patches at the centre; inedible.

## *INOCYBE GEOPHYLLA*

**A** small, shining, white toadstool, with brown gills, and an earthy smell, often found growing in large troops in grassy glades.

*Cap* ⅜–1¼ in/1–3 cm in diameter, conical with a raised centre and an incurved margin, silky white except for a pale ochraceous centre, radially fibrous. *Gills* adnexed, ochraceous to clay-brown, crowded. *Stem* 1¼–2 × ⅛ in/3–5 × 0.3–0.4 cm, cylindrical, silky white, with a slightly swollen base. *Flesh* white, thin but firm, with an earthy smell. *Spore deposit* clay-brown. *Habitat* in troops in both deciduous and coniferous woods; very common. *Edibility* poisonous.

*INOCYBE GEOPHYLLA*

**COMMON WHITE INOCYBE or WHITE FIBER HEAD**

GROUP Mushrooms and Toadstools

FAMILY Cortinarius (Cortinariaceae)

SEASON EDIBILITY

SIMILAR SPECIES

The Lilac Inocybe (*i. geophylla* var. *lilacina*) is identical except for the uniform lilac to violaceous colour, and often grows with the typical variety; poisonous.

## *INOCYBE LANUGINOSA*

**A** dark brown, slender *Inocybe* species, characterized by the erect scales on the cap; grows in damp woodlands.

*Cap* ⅜–2 in/1–5 cm in diameter, convex then flattened, dark brown, radially fibrous, dry, bearing numerous tiny, upturned scales. *Gills* adnexed, cream colour to greyish brown, fairly crowded. *Stem* 1¼–2¾ × ⅛–5⁄16 in/3–7 × 0.3–0.8 cm, tall, cylindrical, brown and scaly. *Flesh* pale, thin. *Spore deposit* clay-brown. *Habitat* among humus in both deciduous and coniferous woodland. *Edibility* poisonous.

*INOCYBE LANUGINOSA*

**WOOLLY INOCYBE**

GROUP Mushrooms and Toadstools

FAMILY Cortinarius (Cortinariaceae)

SEASON EDIBILITY

SIMILAR SPECIES

*I. lanuginella* is paler and less scaly; poisonous.

## *INOCYBE NAPIPES*

**A** small *Inocybe* species, with a central projection on the cap and a distinctly swollen stem base.

*Cap* ¾–1½ in/2–4 cm in diameter, conical then flattened but with a persistent, raised central area, light chestnut-brown, radially fibrous, with the fibres splitting to reveal a white flesh. *Gills* adnexed, ochraceous to clay-brown, crowded. *Stem* 1¼–2¼ × ⅛–3/16 in/3–6 × 0.3–0.5 cm, cylindrical, slender, with a conspicuously swollen base, which sometimes has a rim; similarly coloured to the cap but paler. *Flesh* firm, with a weak smell of meal. *Spore deposit* clay-brown. *Habitat* on the ground in both coniferous and deciduous woodlands. *Edibility* poisonous.

*INOCYBE NAPIPES*

**TURNIP-FOOT INOCYBE**

GROUP Mushrooms and Toadstools

FAMILY Cortinarius (Cortinariaceae)

SEASON EDIBILITY

SIMILAR SPECIES

*I. umbrina* is very similar but with darker brown pigments; poisonous.

## *ROZITES CAPERATA*

**A** large, cream-coloured mushroom, with brown gills, characterized by the conspicuous, membranous ring on the stem, and often occurring in large numbers.

*Cap* 2¼–4¾ in/6–12 cm in diameter, convex then flattened with an upturned, wavy margin, dull straw-yellow, dry, with a hoary-powdery aspect. *Gills* adnate, brown, broad, often vertically wrinkled, crowded. *Stem* 4–6 × ⅜–¾ in/10–15 × 1–2 cm, cylindrical, white to yellowish ochre, with longitudinal striations, and bearing a prominent, white, membranous ring attached to the upper region. *Flesh* thick, white, with a pleasant smell. *Spore deposit* rusty brown. *Habitat* on acidic, sandy soil, under coniferous, sometimes beech, trees, preferring northerly or mountainous localities. *Edibility* said to be a good species after cooking to remove the bitter taste.

*ROZITES CAPERATA*

**GYPSY MUSHROOM**

GROUP Mushrooms and Toadstools

FAMILY Cortinarius (Cortinariaceae)

SEASON EDIBILITY

SIMILAR SPECIES

This species closely resembles a number of the larger *Cortinarius* species, which may be poisonous, but is distinguished by the ring on the stem.

## *CORTINARIUS AMOENOLENS*

**A** beech-wood species, recognized by the yellow scales on a swollen stem-base, the bitter cap cuticle, and the slimy cap.

*Cap* 1¼–5¼ in/3–13 cm in diameter, convex or with a raised centre, pale straw-yellow, slimy, often with whitish scales, near the margin. *Gills* adnate, at first bluish grey and finally light chocolate-brown, crowded. *Stem* 2–5½ × ⅜–¾ in/5–14 × 1–2 cm, cylindrical but with a swollen base (up to 1½ in/4 cm across), which has a narrow rim, at first violaceous blue later developing yellowish tints; yellow patches of the veil present at the base. *Flesh* light blue-violaceous when broken open, with a mild taste except for a bitter cap cuticle, and a pleasant odour of plums. *Spore deposit* cinnamon-brown. *Habitat* beech woods, especially on chalky ground. *Edibility* inedible.

*CORTINARIUS AMOENOLENS*

**FRUITY CORTINARIUS**

GROUP Mushrooms and Toadstools

FAMILY Cortinarius (Cortinariaceae)

SEASON EDIBILITY

SIMILAR SPECIES

*C. calochrous* has a more pronounced margin around the swollen stem base, and a brighter cap colour; inedible.

## *CORTINARIUS COLLINITUS*

**A** typical species of the subgenus *Myxacium,* having a slimy cap and stem, and easily recognized by the bright yellowish brown cap, and conspicuous bands on the stem.

*Cap* 1¼–4 in/3–10 cm in diameter, strongly convex to flattened or with a raised darker centre, tawny orange, slimy. *Gills* adnate, broad, pale violaceous grey becoming rusty brown, crowded. *Stem* 2–4¾ × ³⁄₁₆–¾ in/5–12 × 0.5–2 cm, elongated and cylindrical, off-white to bluish, slimy, with bluish bands of the veil. *Flesh* white, lacking a distinctive odour. *Spore deposit* cinnamon-brown. *Habitat* usually under conifers, sometimes beech trees. *Edibility* inedible.

*CORTINARIUS COLLINITUS*

**SMEARED CORTINARIUS or SLIMY-BANDED CORTINARIUS**

**GROUP** Mushrooms and Toadstools

**FAMILY** Cortinarius (Cortinariaceae)

**SEASON** **EDIBILITY**

**SIMILAR SPECIES**

*C. trivialis* is duller, more slender, and usually grows in wet situations under willow and alder; inedible.

## *CORTINARIUS PURPURASCENS*

**The** cap is slimy but the stem remains dry, characteristic of the subgenus *Phlegmacium*. The purplish to violaceous colour of all parts is distinctive.

*Cap* 2–6 in/5–15 cm in diameter, convex with a raised centre and, often, a wavy edge, very slimy, dark brown, with the margin at first deep violaceous. *Gills* adnate, purplish becoming rusty brown, crowded. *Stem* 2–4¾ × ⅝ -1 in/5–12 × 1.5–2.5 cm, cylindrical with a swollen base, violaceous, bruising darker, with a purplish cortina (veil). *Flesh* pale violaceous, becoming much darker when broken open. *Spore deposit* cinnamon brown. *Habitat* occurs in both deciduous and coniferous woods, often alongside paths. *Edibility* said to be edible, but easily confused with inedible species.

*CORTINARIUS PURPURASCENS*

**PURPLE CORTINARIUS**

**GROUP** Mushrooms and Toadstools

**FAMILY** Cortinarius (Cortinariaceae)

**SEASON** **EDIBILITY**

**SIMILAR SPECIES**

In *C. subpurpurascens* the gills are not violaceous in the young stages; inedible.

## *CORTINARIUS PSEUDOSALOR*

**A** robust, wholly sticky species with a convex, brownish cap and violet tinge on the stem.

*Cap* 1½–3 in/4–8 cm in diameter, slimy, convex or bluntly conical, expanding with a raised centre, greyish brown, often with a violet tinge, slightly grooved at the margin. *Gills* adnate, clay-coloured to cinnamon-brown, not veined. *Stem* 3–4 × ⅜–⅝ in/8–10 × 1–1.5 cm, rather spindle-shaped, white tinted blue-violet above. *Flesh* pale blue-violaceous, with a mild smell of honey, especially when bruised. *Spore deposit* rusty brown. *Habitat* in mixed woodland, fairly common. *Edibility* inedible.

*CORTINARIUS PSEUDOSALOR*

**SHINY COBWEB CAP**

**GROUP** Mushrooms and Toadstools

**FAMILY** Cortinarius (Cortinariaceae)

**SEASON** **EDIBILITY**

**SIMILAR SPECIES**

*C. elatior* is larger, has a more strongly grooved cap margin, pale flesh and the gills are transversely veined; inedible.

## CORTINARIUS PHOLIDEUS

**A** distinctive, brownish species, easily recognized by the shaggy, dark brown scales on the cap and stem, and the violaceous stem apex.

*Cap* 1½–3½ in/4–9 cm in diameter, at first bell-shaped, expanding yet retaining a raised centre, hazel-brown to pale date-brown, darker at centre, covered with darker, raised scales. *Gills* adnate, at first pale violaceous, becoming rusty brown, crowded. *Stem* 2¾–4¾ × ¼–⅜ in/7–12 × 0.7–1 cm, slender, cylindrical, whitish or pale brown, pale violaceous at the top, with a distinct ring-zone below, which is covered with irregular zones of recurved, dark brown, woolly scales. *Flesh* thin, pale brownish, with a violaceous tint at the stem apex. *Spore deposit* rusty brown. *Habitat* in woodland, usually with birch, on damp, peaty soil; uncommon. *Edibility* inedible.

*CORTINARIUS PHOLIDEUS*

**SCALY CORTINARIUS**

**GROUP** Mushrooms and Toadstools

**FAMILY** Cortinarius (Cortinariaceae)

**SEASON** **EDIBILITY**

**SIMILAR SPECIES**

*C. cotoneus* has a velvety cap with an olive-green tint, and the gills are at first yellowish olive; inedible.

## CORTINARIUS SPECIOSISSIMUS

**A** fairly rare species, with a pointed tawny brown cap, growing in coniferous woods.

*Cap* 1¼–3 in/3–8 cm in diameter, convex to conical, with a raised centre, tawny reddish, more yellowish towards the margin, dry, radially fibrous to almost scaly. *Gills* adnate, ochraceous but soon deep tawny brown, rather thick, moderately crowded. *Stem* 2–4¼ × 3⁄16–⅜ in/5–11 × 0.5–1 cm, slightly thickened towards the base, ochre to tawny brown, with a yellow veil forming ring-like zones and patches over the lower region. *Flesh* yellowish, with a faint smell of radish. *Spore deposit* cinnamon-brown. *Habitat* uncommon, in coniferous woodland, among mosses. *Edibility* **deadly poisonous, attacking the liver.**

*CORTINARIUS SPECIOSISSIMUS*

**FOXY ORANGE CORTINARIUS**

**GROUP** Mushrooms and Toadstools

**FAMILY** Cortinarius (Cortinariaceae)

**SEASON** **EDIBILITY**

**SIMILAR SPECIES**

The Poznan Cortinarius (*C. orellanus*) is an equally poisonous species, which has an orange-tawny cap and well-spaced gills, and grows in deciduous woodland. Fortunately, both species are uncommon, but both have been mistaken for chanterelles in the past.

## CORTINARIUS ARMILLATUS

**A** large common species, with distinctive brick-red bands on the stem.

*Cap* 2–4 in/5–10 cm in diameter, fleshy, bell-shaped to almost flattened, moist, orange to tawny brown, smooth but streaky, with an incurved margin. *Gills* adnate, pale brown to dark rusty brown, broad and widely spaced. *Stem* 3–6 × ⅜–⅝ in/8–15 × 1–1.5 cm, tall, with a gradually expanding base (up to 1¼ in/3.5 cm in diameter), solid, brown, with brick-red bands of the veil on the lower surface. *Flesh* whitish, with a smell of radish. *Spore deposit* cinnamon-brown. *Habitat* often with birch, on open heathland, although also known from mixed and coniferous woodland. *Edibility* inedible and best avoided to prevent confusion with deadly poisonous species.

*CORTINARIUS ARMILLATUS*

**RED-BANDED CORTINARIUS or BRACELET CORTINARIUS**

**GROUP** Mushrooms and Toadstools

**FAMILY** Cortinarius (Cortinariaceae)

**SEASON** **EDIBILITY**

**SIMILAR SPECIES**

Unlikely to be confused with other species due to red bands on the stem.

## CORTINARIUS BOLARIS

One of the many *Cortinarius* species with a dry cap and stem, but distinguished by its reddish scales.

*Cap* 1¼–2¾ in/3–7 cm in diameter, convex to depressed, dry, pale yellowish with pinkish to scarlet-red, fibrous scales, bruising red. *Gills* adnate to almost decurrent, coffee colour to rusty brown, crowded. *Stem* 2–3 × 3/16–¾ in/5–8 × 0.5–2 cm, expanded towards the base, which is covered with reddish scales, bruising reddish or ochraceous. *Flesh* brown, turning sulphur-yellow then reddish at the stem base; no odour. *Spore deposit* rusty brown. *Habitat* found solitary in deciduous woodlands, especially with birch and beech. *Edibility* inedible.

*CORTINARIUS BOLARIS*

**RED-DAPPLED CORTINARIUS**

GROUP Mushrooms and Toadstools

FAMILY Cortinarius (Cortinariaceae)

SEASON EDIBILITY

SIMILAR SPECIES

*C. spilomeus* has bluish grey gills and flesh but is otherwise similar; poisonous.

## CORTINARIUS PALEACEUS

A small, fairly common *Cortinarius* species, commonly found, recognized by the delicate white, hair-like scales on the cap and the characteristic smell.

*Cap* ⅜–1¼ in/1–3 cm in diameter, conical and pointed, then expanding, dark brown, with small, curled, white, flaky scales. *Gills* adnate, at first lilac then cinnamon-brown, crowded. *Stem* 1½–2¾ × ⅛ in/4–7 × 0.3–0.5 cm, brown, with several whitish zones. *Flesh* thin, lilac-brown, with a distinctive smell of *Pelargonium*. *Spore deposit* cinnamon-brown. *Habitat* in heather or in damp woods. *Edibility* inedible.

*CORTINARIUS PALEACEUS*

**GERANIUM-SCENTED CORTINARIUS**

GROUP Mushrooms and Toadstools

FAMILY Cortinarius (Cortinariaceae)

SEASON EDIBILITY

SIMILAR SPECIES

The Scurfy Cortinarius (*C. hemitrichus*) lacks the *Pelargonium* smell, and always grows under birch; inedible.

## CORTINARIUS TORVUS

A frequent woodland species which is readily distinguished by the sheathed stem base and comparatively well-developed ring.

*Cap* 1½–4 in/4–10 cm in diameter, convex, date-brown or violet-brown, becoming paler with age, fibrillose at first, becoming smooth. *Gills* adnate, pale violaceous at first becoming rusty brown, broad, rather thick and distant. *Stem* 2¼–4 × ⅜–¾ in/6–10 × 1–2 cm, stout, rather swollen towards the base, pale buff, tinged violet above, the lower part surrounded by a distinct whitish or pale violaceous sheath below a well-marked ring. *Flesh* pale, with violet tinge in the stem and having a sweetish smell. *Spore deposit* rusty brown. *Habitat* in deciduous woods, especially with beech and oak; common. *Edibility* edible.

*CORTINARIUS TORVUS*

**SHEATHED CORTINARIUS**

GROUP Mushrooms and Toadstools

FAMILY Cortinarius (Cortinariaceae)

SEASON EDIBILITY

SIMILAR SPECIES

*C. malachius* is more reddish brown in colour and has different spores.

## *DERMOCYBE SANGUINEA*

A coniferous-wood species recognized by the blood-red colour of cap, gills and stem.

*Cap* ¾–2 in/2–5 cm in diameter, dry, slightly felty, or scaly convex becoming flattened, deep blood-red. *Gills* adnate, crowded, blood-red. *Stem* 2¼–3½ × ⅛–¼ in/6–9 × 0.3–0.6 cm, slender, cylindrical, slightly fibrillose, similarly coloured to the cap, with red cortina, slightly paler at the base. *Flesh* reddish, yielding a reddish juice when squashed, without any distinctive smell. *Spore deposit* rusty brown. *Habitat* in coniferous woods. *Edibility* inedible.

*DERMOCYBE SANGUINEA*

**BLOOD-RED CORTINARIUS**

**GROUP** Mushrooms and Toadstools

**FAMILY** Cortinarius (Cortinariaceae)

**SEASON** **EDIBILITY**

**SIMILAR SPECIES**

*Cortinarius cinnabarinus* is similarly coloured but grows in beech woods; *C. anthracinus* occurs in pine woods but is brownish red and has a cap with a raised centre; *C. phoenicius* is also brownish red in colour and has brown flesh. All three species are poisonous.

## *PHAEOLEPIOTA AUREA*

A large, distinctive species of parks and roadsides, with a powdery, orange-brown cap and striate, funnel-shaped ring.

*Cap* 4–9 in/10–25 cm in diameter, convex, orange-brown, surface scurfy or powdery, the margin often fringed with tooth-like remnants of the veil. *Gills* adnexed, almost free, pale orange-yellow. *Stem* 4–8 × 1¼–1½ in/10–20 × 3–4 cm, stout, pale orange-brown or yellowish brown, sheathed below and bearing a large, membranous ring which is powdery and conspicuously striated on the underside. *Flesh* whitish, without any distinctive smell. *Spore deposit* ochraceous brown. *Habitat* in parks and open woodland, often in clusters. *Edibility* edible, but not recommended.

*PHAEOLEPIOTA AUREA*

**GOLDEN FALSE PHOLIOTA**

**GROUP** Mushrooms and Toadstools

**FAMILY** Cortinarius (Cortinariaceae)

**SEASON** **EDIBILITY**

**SIMILAR SPECIES**

Not likely to be confused with any other specimen.

## *LACTARIUS VELLEREUS*

The largest *Lactarius* species, recognized by the white colour and the very acrid latex.

*Cap* 3–8 in/8–20 cm in diameter, soon depressed to almost funnel-shaped, dry, not zoned, white, with a downy surface and an inrolled margin. *Gills* adnate to decurrent, white to pale cream, thick and moderately distant. *Stem* ¾–3 × ⅜–1¼ in/2–8 × 1–3 cm, white, dry. *Flesh* hard, white, unchanging, with an abundant white latex. *Spore deposit* white. *Habitat* grows in groups in deciduous woodland. *Edibility* inedible, owing to the very acrid taste.

*LACTARIUS VELLERUS*

**FLEECY MILK-CAP**

**GROUP** Mushrooms and Toadstools

**FAMILY** Brittle-gills and Milk-caps (Russulaceae)

**SEASON** **EDIBILITY**

**SIMILAR SPECIES**

The Peppery Milk-cap (*L. piperatus*) has very densely crowded gills; inedible.

## LACTARIUS DELICIOSUS

**A** large species, commonly found in pine woods, forming large troops; and distinguished by the orange colours.

*Cap* 2–8 in/5–20 cm in diameter, convex, soon depressed with an incurved margin, reddish orange with several darker zones, slimy when moist, staining green. *Gills* adnate to short decurrent, pale orange-yellow, staining green, crowded. *Stem* 1¼–3 × ⅜–⅝ in/3–8 × 1–1.5 cm, short, orange, often pitted, staining greenish. *Flesh* thick, yellowish cream, with an orange latex, and a fruity odour. *Spore deposit* pale pinkish cream. *Habitat* often in large numbers under conifers, especially pines. *Edibility* moderately good, with a mild taste.

*LACTARIUS DELICIOSUS*

**SAFFRON MILK-CAP**

**GROUP** Mushrooms and Toadstools

**FAMILY** Brittle-gills and Milk-caps (Russulaceae)

**SEASON** **EDIBILITY**

**SIMILAR SPECIES**

*L. deterrimus* has paler gills, and the stem is not pitted; edible.

## LACTARIUS TORMINOSUS

**The** shaggy, zoned cap with a strongly inrolled margin distinguishes this birchwood species.

*Cap* 2–6 in/5–15 cm in diameter, convex to strongly depressed, pale salmon-pink, with several concentric zones, sticky when wet, with a shaggy, incurved margin. *Gills* adnate to decurrent, pale pinkish, crowded. *Stem* 1¼–3 × 3/16–⅝ in/3–8 × 0.5–1.5 cm, smooth, dry, pinkish but paler than the cap. *Flesh* white to pink, with a white, unchanging latex. *Spore deposit* pale pinkish cream. *Habitat* always with birch, in damp situations. *Edibility* poisonous.

*LACTARIUS TORMINOSUS*

**WOOLLY MILK-CAP or PINK-FRINGED MILK-CAP**

**GROUP** Mushrooms and Toadstools

**FAMILY** Brittle-gills and Milk-caps (Russulaceae)

**SEASON** **EDIBILITY**

**SIMILAR SPECIES**

*L. pubescens* has a whitish cap, while *L. mairei* is orange-coloured; both poisonous.

## LACTARIUS HELVUS

**A** species of damp woodland, recognized by the dry, velvety, buff-coloured cap and watery milk.

*Cap* 2–4¾ in/5–12 cm in diameter, convex or flattened, becoming depressed, often with a small central knob, margin incurved, pale cinnamon or yellowish buff, dry, velvety with tiny, flattened scales. *Gills* adnate or slightly decurrent, pale ochraceous, fairly crowded. *Stem* 2–4¾ × 3/16–1 in/5–12 × 0.5–2.5 cm, cylindrical, often curved, similarly coloured to the cap or slightly darker, smooth or slightly downy. *Flesh* thin, whitish, often hollow in the stem, mild or slightly bitter, smelling of curry when dry; latex almost colourless and slight. *Spore deposit* whitish with pinkish tinge. *Habitat* damp woodland, often with birches or conifers on acid soil. *Edibility* inedible.

*LACTARIUS HELVUS*

**LIQUORICE MILK-CAP**

**GROUP** Mushrooms and Toadstools

**FAMILY** Brittle-gills and Milk-caps (Russulaceae)

**SEASON** **EDIBILITY**

**SIMILAR SPECIES**

*L. azonites* and *L. pterosporus* have dry, slightly velvety caps but differ in colour and milk becoming salmon-pink on cut flesh, and lack of curry-like smell on drying; both inedible.

## *LACTARIUS CHRYSORRHEUS*

**T**he golden yellow latex distinguishes this oak-wood species.

*Cap* 2–4 in/5–10 cm in diameter, convex then expanded to flat, pale yellowish with yellowish pink zones. *Gills* short decurrent, cream-coloured bruising yellow, moderately crowded. *Stem* 2–2¾ × 3⁄16–⅜ in/5–7 × 0.5–1 cm, paler than the cap, smooth. *Flesh* thin, whitish, with a bright golden yellow latex which is mild at first but slowly becomes peppery. *Spore deposit* cream. *Habitat* common in deciduous woodland, especially under oak. *Edibility* inedible.

*LACTARIUS CHRYSORRHEUS*

**YELLOW MILK-CAP**

GROUP Mushrooms and Toadstools

FAMILY Brittle-gills and Milk-caps (Russulaceae)

SEASON EDIBILITY

SIMILAR SPECIES

*L. decipiens* has a smell of *Pelargonium*, and a white latex which turns yellowish after a few seconds; inedible.

## *LACTARIUS TABIDUS*

**A** small, common milk-cap, usually found in damp situations, and distinguished by the mild-tasting latex which discolours yellowish as it dries.

*Cap* ¾–2 in/2–5 cm in diameter, slightly depressed with a raised centre, rusty brown or paler, slightly wrinkled. *Gills* decurrent, pale brown, crowded. *Stem* ¾–1½ × 3⁄16–⅜ in/2–4 × 0.5–1 cm, pale rusty brown. *Flesh* thin, whitish, with a mild to slightly peppery latex. *Spore deposit* pale buff. *Habitat* in deciduous woodland, often growing under birch trees, and among *Sphagnum* moss. *Edibility* inedible.

*LACTARIUS TABIDUS*

**SWEET LACTARIUS**

GROUP Mushrooms and Toadstools

FAMILY Brittle-gills and Milk-caps (Russulaceae)

SEASON EDIBILITY

SIMILAR SPECIES

*L. glyciosmus* has a strong smell of oconut, and *L. theiogalus* is slender with a striated margin; both inedible.

## *LACTARIUS TURPIS*

**A** robust, short-stalked species which has a sticky, dirty-coloured cap, very hot tasting milk, and grows under birches.

*Cap* 2¼–7 in/6–18 cm in diameter, convex, becoming flattened or centrally depressed, incurved at the margin, sticky, but woolly at the margin at first, dark olive-brown or olive-black, sometimes with darker concentric zones, margin paler, yellowish olive. *Gills* narrow, crowded, slightly decurrent, cream or pale buff becoming spotted with brown. *Stem* 1½–3 × ⅜–1 in/4–8 × 1–2.5 cm, short, stout, cylindrical, similarly coloured to the cap or slightly paler, surface sticky. *Flesh* white, hollow in the stem, with abundant white milk; taste very hot and acrid. *Spore deposit* cream. *Habitat* under birches, especially in damp woodland. *Edibility* inedible.

*LACTARIUS TURPIS*

**UGLY MILK-CAP**

GROUP Mushrooms and Toadstools

FAMILY Brittle-gills and Milk-caps (Russulaceae)

SEASON EDIBILITY

SIMILAR SPECIES

Not likely to be confused with any other species.

## LACTARIUS VOLEMUS

**A** large, fleshy, orange-brown milk-cap with abundant white milk and a fishy smell.

*Cap* $2\frac{3}{4}$–$5\frac{1}{4}$ in/7–13 cm in diameter, convex, often depressed at the centre, incurved at the margin, bright orange-brown, smooth or slightly velvety, dry. *Gills* narrow, crowded, slightly decurrent, cream or pale yellowish, staining brown when bruised. *Stem* 2–4 × $\frac{3}{8}$–$\frac{3}{4}$ in/5–10 × 1–2 cm, cylindrical or slightly wider downwards, similarly coloured to the cap, paler near the top, smooth. *Flesh* thick, whitish or pale yellow, yielding copious white milk when cut or broken, and with a distinct fishy smell, especially in older specimens; mild taste. *Spore deposit* white. *Habitat* in mixed woodland and at woodland edge, occasional. *Edibility* a good edible species.

*LACTARIUS VOLEMUS*

**ORANGE-BROWN MILK-CAP or VOLUMINOUS-LATEX MILK-CAP**

GROUP Mushrooms and Toadstools

FAMILY Brittle-gills and Milk-caps (Russulaceae)

SEASON EDIBILITY

SIMILAR SPECIES

Not likely to be confused with any other species. *L. hygrophoroides* is similar in colour, but has distant gills and lacks the fishy odour; edible.

## LACTARIUS RUFUS

**A** commonly found species of coniferous woodland, distinguished by the umbonate, reddish brown cap, and very peppery milk.

*Cap* $1\frac{1}{2}$–4 in/4–10 cm in diameter, convex to flat or depressed with a raised centre, dull bay-red, dry, with an incurved margin. *Gills* short decurrent, white to buffy brown, crowded. *Stem* 2–3 × $\frac{3}{16}$–$\frac{3}{8}$ in/5–8 × 0.5–1 cm, dingy purplish brown, with a white base. *Flesh* fairly thin, pale purplish, with abundant, unchanging white latex. *Spore deposit* pinkish buff. *Habitat* on soil under pines, or in *Sphagnum* bogs. *Edibility* not eaten in Western countries, owing to the acrid taste, yet sold commercially in Finland.

*LACTARIUS RUFUS*

**RUFOUS MILK-CAP or RED-HOT MILK-CAP**

GROUP Mushrooms and Toadstools

FAMILY Brittle-gills and Milk-caps (Russulaceae)

SEASON EDIBILITY

SIMILAR SPECIES

*L. hepaticus* is smaller, has a liver-brown cap, and a latex which discolours yellowish; inedible.

## LACTARIUS CAMPHORATUS

**A** small, reddish brown species, with a distinctive spicy odour, growing as troops in coniferous woodland.

*Cap* $\frac{3}{4}$–2 in/2–5 cm in diameter, broadly convex to depressed, often with a central knob, reddish brown, moist to dry. *Gills* short decurrent, pale purplish brown, narrow, crowded. *Stem* $\frac{3}{4}$–2 × $\frac{3}{16}$–$\frac{3}{8}$ in/2–5 × 0.5–1 cm, cylindrical, reddish brown, smooth. *Flesh* thin, pale brown, with a watery, white, unchanging, mild latex, and a pungent odour. *Spore deposit* yellowish. *Habitat* in coniferous and mixed woodland. *Edibility* spicy odour and taste of dried fruit-bodies is sometimes used as a seasoning.

*LACTARIUS CAMPHORATUS*

**CURRY-SCENTED MILK-CAP or AROMATIC MILK-CAP**

GROUP Mushrooms and Toadstools

FAMILY Brittle-gills and Milk-caps (Russulaceae)

SEASON EDIBILITY

SIMILAR SPECIES

*L. rufus* (see this page) is found in similar situations but is generally larger and lacks the distinctive odour; inedible.

## *LACTARIUS UVIDUS*

The distinguishing characters are the sticky, lilac cap and a white latex which immediately discolours dark lilac-violet.

*Cap* 1½–4 in/4–10 cm in diameter, convex to flattened, pale lilac-brown, smooth, sticky then dry, with the margin incurved at first. *Gills* short decurrent, cream becoming discoloured lilac, crowded. *Stem* 2–3½ × ⅜–⅝ in/5–9 × 1–1.5 cm, off-white, sticky or dry. *Flesh* white, releasing a white, bitter latex which discolours lilac on exposure. *Spore deposit* pale yellow. *Habitat* under birch, more rarely under pine, growing in small groups. *Edibility* said to be poisonous; avoid all species with a purplish-staining latex.

*LACTARIUS UVIDUS*

**MOIST LACTARIUS or COMMON VIOLET-LATEX MILK-CAP**

GROUP Mushrooms and Toadstools

FAMILY Brittle-gills and Milk-caps (Russulaceae)

SEASON EDIBILITY

SIMILAR SPECIES

*L. maculatus*, found in North America, is similar but has a zoned cap.

## *RUSSULA NIGRICANS*

Very common and the largest of the brittle-gills (*Russula* species); recognized by the widely spaced gills and blackening flesh.

*Cap* 4–8 in/10–20 cm in diameter, convex then becoming depressed, fleshy, at first white but soon discolouring dark brown to black, dry. *Gills* adnate, pinkish yellow, finally blackening, very thick and widely spaced. *Stem* 1¼–3 × ⅜–1¼ in/3–8 × 1–3 cm, short and stocky, white then blackening similarly to the cap. *Flesh* thick, white becoming blood-red then black, with a fruity smell. *Spore deposit* white. *Habitat* under various species of trees, often beech. *Edibility* poor.

*RUSSULA NIGRICANS*

**BLACKENING RUSSULA**

GROUP Mushrooms and Toadstools

FAMILY Brittle-gills and Milk-caps (Russulaceae)

SEASON EDIBILITY

SIMILAR SPECIES

*R. adusta* differs in having crowded gills; inedible.

## *RUSSULA DELICA*

A robust, firm, short-stalked species with a strongly inrolled margin, thick, firm flesh and hot, peppery taste.

*Cap* 2¼–6 in/6–15 cm in diameter, convex at first, becoming depressed or funnel-shaped with margin inrolled, dry, slightly downy at first, soon becoming smooth, whitish at first, becoming marked with yellowish brown. *Gills* decurrent, whitish, often bluish near the stem, fairly crowded, often forked, intermediate gills present. *Stem* 1¼–2¼ × ¾–1½ in/3–6 × 2–4 cm, short, stout, white, often with a blue-green zone at the top. *Flesh* white, firm but brittle, not changing colour, with slightly hot taste and distinctive, slightly fishy smell. *Spore deposit* white or pale cream. *Habitat* with deciduous and coniferous trees. *Edibility* edible but worthless.

*RUSSULA DELICA*

**MILK-WHITE RUSSULA**

GROUP Mushrooms and Toadstools

FAMILY Brittle-gills and Milk-caps (Russulaceae)

SEASON EDIBILITY

SIMILAR SPECIES

*R. densifolia* differs in having a sticky cap and flesh turning reddish when broken; inedible. *Lactarius vellereus* is similar in general form and colour, but is readily recognized by the exudation of milk from the broken flesh; inedible.

## *RUSSULA OCHROLEUCA*

**A** common woodland species, recognized by the yellow-ochre cap, white gills, and stem becoming greyish with age.

*Cap* 1½–3½ in/4–9 cm in diameter, convex or flattened, sticky when moist, smooth, yellow-ochre, margin weakly grooved with age. *Gills* adnexed, whitish, rather distant. *Stem* 2–4 × ⅜–¾ in/5–10 × 1–2 cm, cylindrical or slightly wider downwards, veined, white, becoming greyish with age. *Flesh* white, not changing colour, with no distinctive smell and mild or slightly peppery taste. *Spore deposit* white to pale cream. *Habitat* in deciduous and coniferous woodlands. *Edibility* edible when it has been cooked, but poor.

*RUSSULA OCHROLEUCA*

### COMMON YELLOW RUSSULA

**GROUP** Mushrooms and Toadstools

**FAMILY** Brittle-gills and Milk-caps (Russulaceae)

**SEASON** **EDIBILITY**

**SIMILAR SPECIES**

*R. claroflava*, *R. solaris* and *R. lutea* have brighter yellow caps, yellow gills, and deep cream or pale ochre spore deposits. *R. fellea* is straw-coloured throughout and has a fruity smell. All these species are edible, except for *R. solaris* where the edibility is unknown.

## *RUSSULA VIRESCENS*

**A**n uncommon species of deciduous woods, recognized by the green cap which has the surface cracking into small scales.

*Cap* 2¾–4¾ in/7–12 cm in diameter, green or yellowish green, convex, expanding and often depressed at centre, the cuticle towards the margin cracking to expose white flesh and give a scaly appearance. *Gills* almost free, crowded, narrow, white. *Stem* 1½–3½ × ¾–1½ in/4–9 × 2–4 cm, whitish, usually tapered, powdered above, otherwise smooth. *Flesh* white, with no distinctive smell, and mild taste. *Spore deposit* cream. *Habitat* in deciduous woods, often with beech. *Edibility* a good edible species.

*RUSSULA VIRESCENS*

### CRACKED GREEN RUSSULA

**GROUP** Mushrooms and Toadstools

**FAMILY** Brittle-gills and Milk-caps (Russulaceae)

**SEASON** **EDIBILITY**

**SIMILAR SPECIES**

In North America, *R. crustosa* has a more greyish yellow cap, and an ochre spore deposit. *R. aeruginosa*, *R. heterophylla* and *R. cyanoxantha* var. *pelteraui* (see page 55) are greenish, but lack cracking of the cap surface. All these species are edible.

## *RUSSULA XERAMPELINA*

**A**n extremely variable species, but easily recognized by the strong smell of crab.

*Cap* 2–6 in/5–15 cm in diameter, broadly convex to flattened or depressed, very variable in colour, purplish red or paler, sometimes tinted green, with a much darker centre, smooth. *Gills* adnate, yellow to orange-yellow, subdistant, and often interveined. *Stem* 2–4 × ⅜–1¼ in/5–10 × 1–3 cm, enlarged towards the base, white or with a pinkish flush, becoming hollow with age. *Flesh* white, with a mild taste, and an odour of shellfish. *Spore deposit* ochre. *Habitat* under beech or oak trees. *Edibility* good, with a mild, nutty taste.

*RUSSULA XERAMPELINA*

### CRAB RUSSULA

**GROUP** Mushrooms and Toadstools

**FAMILY** Brittle-gills and Milk-caps (Russulaceae)

**SEASON** **EDIBILITY**

**SIMILAR SPECIES**

Many varieties of this species have been described, most based on differences in cap colour.

## *RUSSULA EMETICA*

**A** bright red *Russula* of coniferous woodland, with a very peppery flesh.

*Cap* 1¼–2¼ in/3–6 cm in diameter, hard, convex to flat or with a depressed centre, scarlet red to deep blood-red, smooth, sticky, with an obtusely rounded margin. *Gills* adnexed, off-white, broad, spaced. *Stem* 2–3 × ⅜–¾ in/5–8 × 1–2 cm, pure white, brittle. *Flesh* thick, white, with a strongly acrid taste. *Spore deposit* white. *Habitat* in coniferous woodland. *Edibility* inedible, and can cause vomiting when eaten raw.

*RUSSULA EMETICA*

**EMETIC RUSSULA**

GROUP Mushrooms and Toadstools

FAMILY Brittle-gills and Milk-caps (Russulaceae)

SEASON EDIBILITY

SIMILAR SPECIES

*R. mairei* is a beech-wood species, with a softer flesh; inedible.

## *RUSSULA FRAGILIS*

**A** delicate, purplish *Russula* species, with a grooved cap margin, and a peppery taste.

*Cap* ¾–2¼ in/2–6 cm in diameter, easily broken, convex to depressed, sticky when moist, purplish red or often much paler, with a grooved margin. *Gills* adnexed to adnate, white to pale cream, with an uneven edge. *Stem* ¾–2¾ × 3/16–⅜ in/2–7 × 0.5–1 cm, white, solid, slightly swollen towards the base. *Flesh* brittle, pure white, with a faint fruity odour. *Spore deposit* white. *Habitat* grows commonly in all kinds of woodland. *Edibility* inedible, because of the very acrid taste.

*RUSSULA FRAGILIS*

**FRAGILE RUSSULA**

GROUP Mushrooms and Toadstools

FAMILY Brittle-gills and Milk-caps (Russulaceae)

SEASON EDIBILITY

SIMILAR SPECIES

*R. aquosa* is larger, with a pinkish violet cap, growing in oak woodland, while *R. betularus* is small, pale pink, and grows under birch trees; both inedible.

## *RUSSULA CYANOXANTHA*

**A** common species of deciduous woods, characterized by the variable cap colour, white gills and mild taste.

*Cap* 2–6 in/5–15 cm in diameter, deeply convex becoming expanded and sometimes depressed at the centre, variable in colour, commonly lilac or wine coloured often mixed with olive or greyish yellow, sometimes entirely green (variety *pelteraui*), smooth, sticky when moist. *Gills* adnexed, sometimes forked, flexible, white or very pale cream, fairly crowded. *Stem* 2–3½ × ⅝–1 in/5–9 × 1.5–2.5 cm, white, firm, cylindrical or sometimes wider near the base. *Flesh* white, purplish under the cap cuticle, with no distinctive smell, taste mild. *Spore deposit* white. *Habitat* in deciduous woods. *Edibility* a good edible species.

*RUSSULA CYANOXANTHA*

**BLUE AND YELLOW RUSSULA or CHARCOAL BURNER**

GROUP Mushrooms and Toadstools

FAMILY Brittle-gills and Milk-caps (Russulaceae)

SEASON EDIBILITY

SIMILAR SPECIES

*R. grisea* has a greyish cap, with cream-coloured gills and spore deposit, and a slightly peppery taste; edible.

### *RUSSULA ATROPURPUREA*

**A** woodland species recognized by the reddish purple cap with darker, almost black centre, and white stem.

*Cap* 1½–4¼ in/4–11 cm in diameter, convex, becoming flattened or slightly depressed, deep reddish purple, with centre darker, almost black, developing paler, yellowish blotches, smooth, sticky when moist, margin weakly grooved. *Gills* adnexed, crowded, whitish or cream. *Stem* 1½–3 × ⅝–1 in/4–8 × 1.5–2.5 cm, white at first, becoming greyish with age. *Flesh* thick, firm, white, becoming greyish and spongy in the stem, with slightly fruity smell and taste mild or slightly peppery. *Spore deposit* white. *Habitat* in mixed woodland and parks, especially under oak. *Edibility* edible when cooked, but poor.

*RUSSULA ATROPURPUREA*

**BLACKISH PURPLE RUSSULA**

**GROUP** Mushrooms and Toadstools

**FAMILY** Brittle-gills and Milk-caps (Russulaceae)

**SEASON** **EDIBILITY**

**SIMILAR SPECIES**

*R. fragilis* (see page 55) may have similar cap colour, but is smaller, more fragile, and has jagged gill edges and a hot, peppery taste; inedible.

### *HYGROPHOROPSIS AURANTIACA*

**A** distinctive, pine-wood toadstool recognized by the orange colours and thin, crowded, much-forked, decurrent gills.

*Cap* 1–3 in/2.5–8 cm in diameter, convex at first, later depressed, incurved at the margin, orange-yellow (whitish in variety *pallida*), soft, dry, slightly felty. *Gills* decurrent, deep orange (cream in variety *pallida*), thin, crowded, much forked. *Stem* 1–2¾ × 3⁄16–⅜ in/2.5–7 × 0.4–1 cm, cylindrical or tapered, yellowish or orange-brown, often curved. *Flesh* yellowish (whitish in variety *pallida*). *Spore deposit* white. *Habitat* in pine woods and on heaths, often in troops. *Edibility* inedible.

*HYGROPHOROPSIS AURANTIACA*

**FALSE CHANTERELLE**

**GROUP** Boletes

**FAMILY** Paxillus (Paxillaceae)

**SEASON** **EDIBILITY**

**SIMILAR SPECIES**

Often confused with The Chanterelle (*Cantharellus cibarius*, page 19), but has thin gills and is orange rather than ochre-yellow; edible.

### *PAXILLUS INVOLUTUS*

**A** very common, large species with a short stem, and brown gills which feel slimy when squashed.

*Cap* 1½–8 in/4–20 cm in diameter, convex to flat, soon depressed, brown, dry although slimy when wet, with a ridged margin which remains inrolled. *Gills* decurrent, olive-green – yellowish, bruising brown, forked and crowded, and easily detached from the cap. *Stem* 1½–2¾ × ⅜–¾ in/4–7 × 1–2 cm, firm, yellowish brown, bruising reddish brown, smooth. *Flesh* yellowish, thick, soft. *Spore deposit* clay-brown. *Habitat* on the ground in mixed woods or heathland, often with birch. *Edibility* poisonous; thought to have an accumulative toxin and therefore to be avoided.

*PAXILLUS INVOLUTUS*

**BROWN ROLL-RIM or POISON PAXILLUS**

**GROUP** Boletes

**FAMILY** Paxillus (Paxillaceae)

**SEASON** **EDIBILITY**

**SIMILAR SPECIES**

*P. rubicundulus* has a scaly cap and yellowish gills, while *P. filamentosus* does not bruise reddish brown; both poisonous.

## PAXILLUS ATROTOMENTOSUS

**A** large mushroom which is found on conifer stumps, and is recognized by its dark, furry stem.

*Cap* 2–12 in/5–30 cm in diameter, convex to depressed, dry, yellowish to chestnut-brown, felty, with the margin remaining inrolled. *Gills* decurrent, yellowish ochre, spotted brown, narrow, forked, very crowded, easily detached from the cap. *Stem* 1½–3½ × ¾–1½ in/4–9 × 2–4 cm, short and stout, often off-central to almost laterally attached, with a blackish brown, velvety surface. *Flesh* tough, thick, cream colour. *Spore deposit* yellowish brown. *Habitat* around pine and spruce stumps. *Edibility* inedible, unpalatable with a bitter taste.

*PAXILLUS ATROTOMENTOSUS*

**DARK DOWNY PAXILLUS or VELVET-FOOTED PAXILLUS**

GROUP Boletes

FAMILY Paxillus (Paxillaceae)

SEASON EDIBILITY

SIMILAR SPECIES

Other *Paxillus* species lack the furry stem.

## PHYLLOPORUS RHODOXANTHUS

**A** fairly rare species, resembling a *Xerocomus* bolete (see next two species), but with bright yellow, thick gills.

*Cap* ¾–2¾ in/2–7 cm in diameter, convex then expanding, dry, reddish yellow to reddish brown, velvety and smooth. *Gills* deeply decurrent, bright yellow, bruising blue, broad, waxy, thick, at times forked and interveined, widely spaced, becoming pore-like over the stem apex. *Stem* ¾–3½ × 3/16–⅝ in/2–9 × 0.5–1.5 cm, tapering to the base, reddish brown, ridged, solid. *Flesh* thick, yellow. *Spore deposit* yellowish green. *Habitat* on poor soil, usually under conifers. *Edibility* said to be edible after cooking.

*PHYLLOPORUS RHODOXANTHUS*

**GILLED BOLETE**

GROUP Boletes

FAMILY Xerocomus (Xerocomaceae)

SEASON EDIBILITY

SIMILAR SPECIES

This species closely resembles the edible *Xerocomus subtomentosus* but has gills instead of pores and tubes.

## XEROCOMUS CHRYSENTERON

**A** common but variable bolete, often found near to towns; the velvety cap with reddish cracks and the yellow pores are distinctive.

*Cap* 1½–3 in/4–8 cm in diameter, convex to flat, dry, olive-brown to reddish brown, velvety when young, cracking to reveal underlying red flesh. *Tubes* slightly depressed around the stem apex, bright yellow but slowly discolouring blue-green; *pores* angular, bright yellow, staining greenish. *Stem* 1½–3 × 3/16–⅝ in/4–8 × 0.5–1.5 cm, slender, scurfy, yellowish with pinkish red tints, solid. *Flesh* thick, white to yellowish, slowly discolouring blue when broken. *Spore deposit* cinnamon-brown. *Habitat* in deciduous woodland, especially under oak trees. *Edibility* edible but not recommended.

*XEROCOMUS CHRYSENTERON*

**RED-CRACKED BOLETUS**

GROUP Boletes

FAMILY Xerocomus (Xerocomaceae)

SEASON EDIBILITY

SIMILAR SPECIES

*X. subtomentosus* is very similar but lacks the red cracking on the cap surface; edible.

## *XEROCOMUS BADIUS*

**A** coniferous-wood species distinguished by its bay-brown, sticky cap and yellowish pores, which stain rapidly blue when they are bruised.

*Cap* 2–4¾ in/5–12 cm across, hemispherical, sometimes becoming flattened, deep bay-brown, smooth, slightly sticky when moist. *Tubes* adnate, pale yellow; *pores* lemon-yellow or olive-yellow, angular, rapidly turning blue-green when bruised. *Stem* 2–4 × ⅜–1¼ in/5–10 × 1–3 cm, cylindrical, brown, paler than the cap, with fine but distinct darker longitudinal streaks. *Flesh* thick, whitish or pale yellowish, becoming faintly blue when cut or broken, especially in the cap over the tubes. *Spore deposit* olive-brown. *Habitat* in woods, usually with conifers. *Edibility* a good edible species.

*XEROCOMUS BADIUS*

**BAY BOLETUS**

GROUP Boletes

FAMILY Xerocomus (Xerocomaceae)

SEASON EDIBILITY

SIMILAR SPECIES

*Gyroporus castaneus* (see page 60) may be superficially similar, but has white flesh and does not bruise blue; edible.

## *BOLETUS EDULIS*

**A** robust species recognized by the brown cap, pale stem with white network on the upper part, and white, unchanging flesh.

*Cap* 3–8 in/8–20 cm across, usually hemispherical, brown, slightly paler at the margin, smooth, dry or slightly sticky in moist weather. *Tubes* white or cream; *pores* whitish, small. *Flesh* whitish, not changing colour when cut. *Stem* 3–7 × 1½–2¾ in/8–18 × 4–7 cm, thickened below, whitish or pale brown, bearing in the upper part a network of white, raised lines. *Spore deposit* olive-brown. *Habitat* common in all types of woodland, mostly beech and oak, occasionally pine. *Edibility* a good edible species, much sought after and used commercially as a flavouring in soups.

*BOLETUS EDULIS*

**CEP or PENNY BUN BOLETUS**

GROUP Boletes

FAMILY Boletus (Boletaceae)

SEASON EDIBILITY

SIMILAR SPECIES

*B. aestivalis* is very similar, but has a paler cap and a coarse network covering the whole length of the stem; edible.

## *BOLETUS ERYTHROPUS*

**A** large bolete characterized by the brown cap, red pores, red-dotted stem, and flesh which turns instantly bright blue when cut.

*Cap* 3–8 in/8–20 cm across, bay-brown or reddish brown, hemispherical at first, expanding somewhat, finely downy becoming smooth, slightly sticky when moist. *Tubes* lemon-yellow or greenish, becoming dark blue when cut; *pores* small, round, orange-red, becoming dark blue when bruised. *Stem* 2–5½ × ¾–2 in/5–14 × 2–5 cm, robust, usually thickened downwards, yellowish, but densely covered with red dots. *Flesh* yellowish, becoming instantly dark blue when cut or broken, lacking a distinctive smell. *Spore deposit* olive-brown. *Habitat* in mixed woods. *Edibility* edible when cooked; may cause stomach upsets if eaten raw.

*BOLETUS ERYTHROPUS*

**DOTTED STEM BOLETUS**

GROUP Boletes

FAMILY Boletus (Boletaceae)

SEASON EDIBILITY

SIMILAR SPECIES

*B. luridus* has the stem covered with a red, raised network. *B. dupainii* has a red-dotted stem, but has a red cap. Both species are poisonous.

## BOLETUS SATANAS

**A** large and distinctive species having a whitish cap, red pores and a distinct red network on the stem.

*Cap* 3–9 in/8–25 cm across, pale, whitish or with buff tinge, bruising brown when touched, soon becoming smooth, cracking slightly with age. *Tubes* olive-yellow, turning blue when cut; *pores* small, round, red, bruising greenish blue. *Stem* 2¼–3½ × 2–4¾ in/6–9 × 5–12 cm, short, stout, swollen at the base, yellowish at the top, reddish below, covered with a distinct red network of raised lines. *Flesh* pale straw or whitish, slowly turning pale blue when cut or broken and often with rusty patches in the stem; smell unpleasant. *Spore deposit* olive-brown. *Habitat* in deciduous woodland, especially with oak and beech, usually on chalky soil; rare. *Edibility* poisonous.

*BOLETUS SATANAS*

**DEVIL'S BOLETUS**

GROUP Boletes

FAMILY Boletus (Boletaceae)

SEASON EDIBILITY

SIMILAR SPECIES

*B. satanoides* differs in having a pinkish cap, less swollen stem and network developed only on the upper stem; poisonous.

## TYLOPILUS FELLEUS

**A** large bolete, characterized by the dark brown network over the stem surface, pinkish pores, and an instantly bitter taste.

*Cap* 2–6 in/5–15 cm in diameter, convex becoming flattened, dark brown to buffy brown, dry, sticky when wet, with an obtuse margin. *Tubes* adnate, depressed around the stem apex, white to pinkish, bruising brown; *pores* minute, pinkish white, bruising brown. *Stem* 1½–4 × ¾–1¼ in/4–10 × 2–3 cm, club-shaped with a swollen base, pale brown, with a dark brown, net-like pattern. *Flesh* thick, white, unchanging. *Spore deposit* clay-pink. *Habitat* under beech and oak. *Edibility* inedible owing to extremely bitter taste, which will ruin any dish.

*TYLOPILUS FELLEUS*

**BITTER BOLETUS**

GROUP Boletes

FAMILY Boletus (Boletaceae)

SEASON EDIBILITY

SIMILAR SPECIES

The edible *Boletus edulis* (see page 58) and relatives have a white net-like pattern on the stem and are not bitter.

## LECCINUM SCABRUM

**A** very common species, belonging to a group sometimes known as the "rough shanks" owing to their scaly stems. This species has a greyish brown cap and blackish stem scales.

*Cap* 1½–4 in/4–10 cm in diameter, strongly convex, thick fleshed, greyish brown to yellowish brown, smooth, dry to sticky. *Tubes* adnexed, deeply sunken around the stem apex, off-white; *pores* off-white, minute, bruising brownish. *Stem* 2¾–6 × ¾–1¼ in/7–15 × 2–3 cm, tall, off-white, with tiny, black, granular scales. *Flesh* thick, soft, white, unchanging or with slight pink-brown flush. *Spore deposit* cinnamon-brown. *Habitat* on the ground, under birch trees. *Edibility* edible and excellent.

*LECCINUM SCABRUM*

**BROWN BIRCH BOLETE or COMMON SCABER STALK**

GROUP Boletes

FAMILY Boletus (Boletaceae)

SEASON EDIBILITY

SIMILAR SPECIES

*L. holopus* has a white cap and grows with birch, while *L. vulpinus* is purplish brown and grows with pine; both edible.

## *LECCINUM VERSIPELLE*

Commonly found in birch woods, this species is characterized by a distinctly orange cap and a blackening flesh.

*Cap* 3–6 in/8–15 cm in diameter, strongly convex, orange to yellowish brown, dry, smooth but with a shaggy margin. *Tubes* adnate, sunken around the stem apex, greyish; *pores* minute, pale grey. *Stem* 4–6 × ¾–1¼ in/10–15 × 2–3 cm, tall, usually with a slightly swollen base, white, covered with small, black, granular scales. *Flesh* thick, white then discolouring black when exposed to air. *Spore deposit* cinnamon-brown. *Habitat* numerous on the ground in birch woods. *Edibility* edible and good.

*LECCINUM VERSIPELLE*

**ORANGE BIRCH BOLETE**

GROUP Boletes

FAMILY Boletus (Boletaceae)

SEASON EDIBILITY

SIMILAR SPECIES

*L. aurantiacum* has orange-brown scales on the stem, and grows with poplar; edible.

## *GYROPORUS CASTANEUS*

A fairly uncommon bolete, which can be recognized by the hollow stem that is often transversely cracked.

*Cap* 1¼–4 in/3–10 cm in diameter, broadly convex to flat; chestnut-brown to tawny, dry. *Tubes* adnexed, sunken around the stem apex, white to lemon-yellow; *pores* small, round to angular, white then yellowish, bruising brown. *Stem* 1½–2¾ × 3⁄16–⅜ in/4–7 × 0.5–1 cm, pale brown, tapering above and below, finally hollow, with a ridged or cracked surface. *Flesh* white, unchanging or pinkish. *Spore deposit* yellow. *Habitat* on the ground under oak, beech and sweet chestnut. *Edibility* edible and excellent, apart from the tough stems.

*GYROPORUS CASTANEUS*

**CHESTNUT BOLETE**

GROUP Boletes

FAMILY Boletus (Boletaceae)

SEASON EDIBILITY

SIMILAR SPECIES

The Bluing Bolete (*G. cyaneus*) has a paler brown cap, and all parts bruise deep blue; edible.

## *SUILLUS GRANULATUS*

A pine-wood species recognized by the slimy, orange-brown cap, yellow pores which exude milky drops, and yellowish granules at the top of the stem.

*Cap* 1½–4 in/4–10 cm across, convex, orange-brown or reddish brown, becoming yellowish orange with age. *Tubes* adnate, pale yellow, not changing colour when cut; *pores* small, yellow, often weeping watery droplets. *Stem* 1½–3½ × ¼–½ in/4–9 × 0.7–1.2 cm, slightly tapered, pale lemon-yellow, the upper part covered with whitish or pale yellow granules which exude milky droplets. *Flesh* thick, soft, pale lemon-yellow. *Spore deposit* ochraceous. *Habitat* under conifers, especially pines. *Edibility* edible but poor.

*SUILLUS GRANULATUS*

**GRANULATED BOLETUS**

GROUP Boletes

FAMILY Boletus (Boletaceae)

SEASON EDIBILITY

SIMILAR SPECIES

*S. collinitus* has a brown cap, and a stem bearing bright red-brown dots; edible.

## *SUILLUS LUTEUS*

**A** pine-wood species that is characterized by the very slimy, chestnut cap and well-developed ring on the stem.

*Cap* 2–4¾ in/5–12 cm across, convex, very slimy, dull chestnut-brown or chocolate-brown with purplish tinge. *Tubes* adnate, pale yellow; *pores* small, lemon-yellow or straw coloured. *Stem* 2–3 × ¾–1¼ in/5–8 × 2–3 cm, bearing a large ring, brownish below, whitish or pale yellowish above, with brownish, glandular dots; *ring* spreading, cream or yellowish, membranaceous but partly rubbed away in age. *Flesh* white, not changing colour when cut, without a distinctive smell. *Spore deposit* clay or olivaceous. *Habitat* under conifers, especially pines. *Edibility* edible but poor.

*SUILLUS LUTEUS*

### SLIPPERY JACK

GROUP Boletes

FAMILY Boletus (Boletaceae)

SEASON EDIBILITY

SIMILAR SPECIES

*S. grevillei* has a less conspicuous ring, a chrome-yellow cap, and occurs under larch; edible.

## *STROBILOMYCES STROBILACEUS*

**A** very distinctive species, readily recognized by the grey-brown colours, scaly cap and stem, and flesh becoming reddish when it has been cut.

*Cap* 3–8 in/8–20 cm across, hemispherical, expanding somewhat, covered with large, woolly, grey-brown or mouse-grey scales, margin shaggy with tooth-like remains of the veil. *Tubes* whitish or pale grey, reddening when cut; *pores* rather large, angular, greyish. *Stem* 2¾–4¾ × ⅝–1 in/7–12 × 1.5–2.5 cm, cylindrical, similarly coloured to the cap, paler above, somewhat scaly, bearing a shaggy ring. *Flesh* pale greyish, becoming reddish and finally greyish black when cut, without a distinctive smell. *Spore deposit* violaceous black. *Habitat* in mixed woodland, especially in mature beech woods, also with oak and pine. *Edibility* edible but worthless.

*STROBILOMYCES FLOCCOPUS*

### OLD MAN OF THE WOODS

GROUP Boletes

FAMILY Boletus (Boletaceae)

SEASON EDIBILITY

SIMILAR SPECIES

Not likely to be confused with any other species.

## *CHROOGOMPHUS RUTILUS*

**A** fairly common species of pine woods, recognized by the dark, thick, decurrent gills and the bright yellow stem base.

*Cap* 1¼–4 in/3–10 cm in diameter, convex with a central knob, reddish brown, sticky, smooth, with an incurved margin. *Gills* decurrent, ochraceous soon darkening to deep purplish brown, thick and broad, often forking, well spaced. *Stem* 2–4¾ × 3⁄16–⅜ in/5–12 × 0.5–1 cm, tapering towards the base, ochraceous but developing a pinkish tint at the apex and chrome yellow at the base, solid; bearing a cobweb-like ring. *Flesh* pinkish to yellow, firm. *Spore deposit* blackish brown. *Habitat* under pines, usually among moss. *Edibility* edible.

*CHROOGOMPHUS RUTILUS*

### PINE SPIKE CAP

GROUP Boletes

FAMILY Spike Cap (Gomphidiaceae)

SEASON EDIBILITY

SIMILAR SPECIES

*C. vinicolor* has a wine-red cap and yellowish gills, and the Slimy Spike Cap (*Gomphidius glutinosus*) has a slimy veil and is usually found with spruce trees; both edible.

## GOMPHIDIUS ROSEUS

One of the smaller "spike caps", easily distinguished by the rosy colour of the cap.

*Cap* $\frac{3}{4}$–2 in/2–5 cm in diameter, convex then flattened, rose-red, smooth, slimy, with an incurved margin. *Gills* decurrent, at first pale then dark greyish brown, forking, thick and well spaced. *Stem* $1\frac{1}{4}$–$1\frac{1}{2}$ × $\frac{3}{16}$–$\frac{5}{8}$ in/3–4 × 0.5–1.5 cm, tapering to a point at the base, solid, whitish but covered with a pale pinkish to orange, slimy veil, and provided with a poorly defined ring-zone. *Flesh* whitish to grey, soft. *Spore deposit* blackish brown. *Habitat* usually under pine trees, and nearly always associated with the bolete, *Suillus bovinus*. *Edibility* edible.

*GOMPHIDIUS ROSEUS*

### ROSY SPIKE CAP

**GROUP** Boletes

**FAMILY** Spike Cap (Gomphidiaceae)

**SEASON** **EDIBILITY**

**SIMILAR SPECIES**

*G. glutinosus* has a grey-brown slimy cap, and the stem base is bright yellow; edible.

## CLAVARIADELPHUS PISTILLARIS

The large, simple, club-shaped fruitbodies are unlikely to be confused with those of any other species.

*Fruitbody* 4–9 in/10–25 cm tall and up to 6 cm in diameter, swollen club-shaped with a rounded apex and tapering below into a slender stem, ochre-yellow, progressively discolouring reddish brown towards the base, smooth or with a few vertical wrinkles. *Flesh* thick, firm, white bruising purplish brown. *Spore deposit* white. *Habitat* among leaf-litter in beech woods, or sometimes found associated with pines at higher localities. *Edibility* edible after cooking.

*CLAVARIADELPHUS PISTILLARIS*

### GIANT CLUB or PESTLE-SHAPED CORAL

**GROUP** Club Fungi

**FAMILY** Giant Club (Clavariadelphaceae)

**SEASON** **EDIBILITY**

**SIMILAR SPECIES**

The flat-topped Club (*C. truncatus*) has a flattened, sterile apex; edible. *C. fistulosus* is tall and slender, rarely more than $\frac{3}{16}$ in/ 0.5 cm in diameter; inedible.

## CLAVULINA CRISTATA

A very common coral fungus, with numerous, white branches.

*Fruitbody* $\frac{3}{4}$–$2\frac{3}{4}$ in/2–7 cm in diameter, densely branched, white, with the individual branches terminating as crest-like tips; arising from an indistinct stalk-like base. *Flesh* white, soft. *Spore deposit* white. *Habitat* on the ground in deciduous and mixed woodland. *Edibility* edible but poor.

*CLAVULINA CRISTATA*

### CRESTED CORAL FUNGUS

**GROUP** Club Fungi

**FAMILY** Clavulina (Clavulinaceae)

**SEASON** **EDIBILITY**

**SIMILAR SPECIES**

The Grey Coral Fungus (*C. cinerea*) is very closely related but coloured ash-grey; edible.

## *CLAVULINA RUGOSA*

Commonly found in mixed woodland, and recognized by the irregular, thick and wrinkled whitish clubs.

*Fruitbody* 2–3 in/5–8 cm tall, forming one or more clubs, 0.5–1 cm in diameter, whitish or with a yellowish tint, flattened, with a few vertical wrinkles or ridges, and a blunt, rounded apex. *Flesh* whitish, fragile. *Spore deposit* white. *Habitat* on bare soil in mixed woodland. *Edibility* edible but poor.

*CLAVULINA RUGOSA*

**WRINKLED CLUB**

**GROUP** Club Fungi

**FAMILY** Clavulina (Clavulinaceae)

**SEASON** **EDIBILITY**

**SIMILAR SPECIES**

The other *Clavulina* species are highly branched.

## *RAMARIA BOTRYTIS*

Only occasionally found, this large white coral fungus may be recognized by the reddish tip on each of the many branches.

*Fruitbody* 1¼–6 in/3–15 cm in diameter, rather cauliflower-like, with a stout basal stalk which gives rise to numerous short branches which themselves frequently branch, white to pale yellowish, forming spiny, wine-red tips. *Flesh* thin, white, unchanging. *Spore deposit* pale brown. *Habitat* solitary on soil, especially in beech woods but sometimes also under conifers. *Edibility* edible, apart from the bitter tips.

*RAMARIA BOTRYTIS*

**PURPLE-TIPPED CORAL FUNGUS or CLUSTERED CORAL FUNGUS**

**GROUP** Club Fungi

**FAMILY** Coral Fungi (Ramariaceae)

**SEASON** **EDIBILITY**

**SIMILAR SPECIES**

Not to be confused with the poisonous *R. formosa* (see this page), which is salmon-pink, and causes stomach upsets.

## *RAMARIA FORMOSA*

An occasional species that forms large, salmon-pink tufts, with yellow tips.

*Fruitbody* 2–8 in/5–20 cm in diameter, with dense branching arising from a short but stout basal stalk, salmon-pink with pale yellowish tips, bruising brown. *Flesh* fibrous, pinky brown. *Spore deposit* pale brown. *Habitat* on the ground under conifers. *Edibility* poisonous, known to cause severe stomach upsets.

*RAMARIA FORMOSA*

**YELLOW-TIPPED CORAL FUNGUS**

**GROUP** Club Fungi

**FAMILY** Coral Fungi (Ramariaceae)

**SEASON** **EDIBILITY**

**SIMILAR SPECIES**

*R. aurea* does not stain brown on bruising and *R. botrytis* (see this page) is whitish with red-tipped branches; both poisonous.

## *COLTRICIA PERENNIS*

**A** polypore with a brown and zoned funnel-cap and slender, central stem.

*Cap* ¾–4 in/2–10 cm in diameter, circular with a depressed centre to almost funnel-shaped, golden brown to cinnamon-brown, finely velvety, with numerous concentric zones, and a thin, straight margin. *Tubes* decurrent, reddish brown; *pores* small, angular, rusty brown. *Stem* 1¼–2¾ × 3/16–⅜ in/3–7 × 0.5–1 cm, cylindrical, similarly coloured to the cap or darker, velvety, solid. *Flesh* firm, fibrous, brown. *Habitat* on sandy soil under conifers. *Edibility* inedible.

*COLTRICIA PERENNIS*

**BROWN FUNNEL POLYPORE**

GROUP Bracket Fungi

FAMILY Brown Polypore (Hymenochaetaceae)

SEASON EDIBILITY

SIMILAR SPECIES

The Cinnamon Polypore (*C. cinnamomea*) is smaller, has a shiny cap, and grows in groups in deciduous woodland; inedible.

## *POLYPORUS UMBELLATUS*

**A** rare but distinctive species, forming a compound fruitbody which has numerous small caps originating from an underground sclerotium.

*Compound fruitbody* up to 20 in/50 cm in diameter, bearing numerous centrally attached caps on slender stems. *Caps* ⅜–1¼ in/1–3 cm in diameter, circular and flattened, yellowish brown, with minute, flattened scales, otherwise smooth, with a thin margin. *Tubes* decurrent, white, firm; *pores* angular or elongate, small, white. *Stem* thick at the base, up to 3 cm in diameter, much branched, white to straw-yellow, smooth, solid, arising from a large, underground, irregularly branching, tuber-like sclerotium, which has a blackish brown crust. *Flesh* thin, white, fibrous. *Spore deposit* white. *Habitat* on the ground, near deciduous trees. *Edibility* said to be edible.

*POLYPORUS UMBELLATUS*

**UMBRELLA POLYPORE**

GROUP Bracket Fungi

FAMILY True Polyporus (Polyporaceae)

SEASON EDIBILITY

SIMILAR SPECIES

*Meripilus giganteus* and *Grifola frondosa* (for both see page 113) have laterally attached, not central, caps, and their fruit-bodies grow directly on dead wood; both inedible.

## *THELEPHORA TERRESTRIS*

**C**ommonly found in groups, among fallen conifer needles.

*Fruitbody* forming rosette-like clusters, sometimes more upright and vase-shaped, ¾–2¼ in/2–6 cm in diameter, consisting of several fan-shaped lobes with an irregular thin margin; upper surface greyish brown or darker, radially fibrous to scaly, with indistinct zoning; lower surface smooth or wrinkled, cocoa-brown. *Flesh* thin, tough and leathery, fibrous, with an earthy odour. *Spore deposit* cocoa-brown. *Habitat* on the ground, in coniferous woodland. *Edibility* inedible.

*THELEPHORA TERRESTRIS*

**EARTH FAN or COMMON FIBER VASE**

GROUP Bracket Fungi

FAMILY Earth Fan (Thelephoraceae)

SEASON EDIBILITY

SIMILAR SPECIES

*T. anthocephala* has erect, narrow branches, often with a whitish margin; inedible.

## *GEASTRUM TRIPLEX*

One of the largest earthstars, with thick, star-like, fleshy rays, and a collar around the spore-sac.

*Fruitbody* up to 2¼ in/6 cm in diameter, in the young, unopened state resembling a tulip bulb. *Spore-sac* ¾–1½ in/2–4 cm in diameter, globose, greyish brown, containing a powdery spore-mass, with an apical pore and surrounded by a thick, fleshy collar. *Rays* 5–6, thick and fleshy, curved downwards, creamy brown, often cracked on the upper surface, greyish on the underside. *Spore deposit* dark brown at maturity. *Habitat* in groups among leaf-litter, under beech, preferring chalky soil. *Edibility* inedible.

*GEASTRUM TRIPLEX*

**COLLARED EARTHSTAR**

**GROUP** Puffballs and allies

**FAMILY** Earthstar (Geastraceae)

**SEASON** **EDIBILITY**

**SIMILAR SPECIES**
The Beaked Earthstar (*G. pectinatum*) is much thinner and smaller, and the spore-sac is borne on a short stalk, while the Rounded Earthstar (*G. saccatum*) lacks both a stalk and a collar; both inedible.

## *GEASTRUM FORNICATUM*

A rare earthstar in which the rays split and become strongly arched.

*Fruitbody* up to 3 in/8 cm in diameter, and up to 4 in/10 cm high. *Spore-sac* globose, brown, with a torn apical pore and supported by a short stalk. *Rays* 4, brown, with the inner layer separating and rising above the outer layer, but remaining attached at the tips. *Spore deposit* blackish brown at maturity. *Habitat* among litter in deciduous woods. *Edibility* inedible.

*GEASTRUM FORNICATUM*

**ARCHED EARTHSTAR**

**GROUP** Puffballs and allies

**FAMILY** Earthstar (Geastraceae)

**SEASON** **EDIBILITY**

**SIMILAR SPECIES**
*G. quadrifidum* is more slender, and the spore-sac pore is surrounded by a circular groove; inedible.

## *CALVATIA EXCIPULIFORMIS*

A large, pestle-shaped puffball recognized by the well-developed stem, and upper part breaking open irregularly to expose the spore-mass.

*Fruitbody* 3–8 in/8–20 cm high, club-shaped or pestle-shaped, head rounded, 1¼–4 in/3–10 cm in diameter, somewhat furrowed beneath, narrowed below to a tall stem-like base, wholly whitish or pale buff, surface at first bearing groups of small, brownish spines united at their tips, soon becoming smooth. Wall papery at maturity, breaking open irregularly at the top to expose the spores. *Flesh* whitish throughout at first, the fertile part becoming yellowish brown and then olive-brown as the spores develop. *Sterile base* firm, spongy, pale brown. *Habitat* in mixed woodland, on heaths and waste ground. *Edibility* edible when young and still white inside.

*CALVATIA EXCIPULIFORMIS*

**PESTLE-SHAPED PUFFBALL**

**GROUP** Puffballs and allies

**FAMILY** Puffball (Lycoperdaceae)

**SEASON** **EDIBILITY**

**SIMILAR SPECIES**
Not likely to be confused with any other species.

## LYCOPERDON PERLATUM

A common species characterized by the surface which bears numerous conical spines, and in opening by a pore at the top.

*Fruitbody* 1¼–3 in/3–8 cm high, ¾–2¼ in/2–6 cm in diameter, club-shaped, with a stem-like base, upper part rounded and often somewhat raised at the centre, whitish at first, finally pale brown, opening by a small pore at the top. *Outer layer* composed of short, pyramidal spines each surrounded by a ring of smaller warts, leaving a somewhat net-like pattern when rubbed. *Inner tissue* white, becoming olive-brown as the spores develop. *Sterile base* firm, spongy, composed of small cells. *Habitat* woodland and pastures. *Edibility* edible when still white inside.

*LYCOPERDON PERLATUM*

**COMMON PUFFBALL or GEM-STUDDED PUFFBALL**

GROUP Puffballs and allies

FAMILY Puffball (Lycoperdaceae)

SEASON EDIBILITY

SIMILAR SPECIES

*L. foetidum* is darker in colour, with groups of blackish brown spines which are convergent at their tips. *L. pedicellatum* differs in having spines 1/16 in/0.1–0.2 cm long, often in groups convergent at their tips. Both species are edible.

## SCLERODERMA CITRINUM

A yellowish, species which lacks a stalk and has a thick, scaly wall.

*Fruitbody* 1½–4 in/4–10 cm across, subglobose, lacking a stalk, arising from white, string-like mycelium, yellowish or yellowish brown, the surface splitting into thick, conspicuous scales. *Wall* thick, breaking open irregularly at maturity to expose the spore mass. *Inner tissue* whitish at first, soon purplish black patterned with white veins, firm, with unpleasant smell, finally powdery when the spores are mature. *Habitat* on heaths and in open woodland, especially on sandy soil. *Edibility* inedible.

*SCLERODERMA CITRINUM*

**COMMON EARTHBALL or PIGSKIN POISON PUFFBALL**

GROUP Puffballs and allies

FAMILY Earthball (Sclerodermataceae)

SEASON EDIBILITY

SIMILAR SPECIES

*S. geaster* is similar, but is brown and opens by spreading star-like rays; inedible.

## CLATHRUS RUBER

A rare species, confined to warmer regions, but easily recognized by the bright red, spherical net, arising out of a white "egg".

*Fruitbody* originates as an off-white egg, which is ¾–1¼ in/2–3 cm in diameter, rather gelatinous, and breaks open to release a globose network of bright red to salmon-pink, hollow arms, bearing a green, slimy, spore-bearing structure (gleba) on the inner surfaces. *Spore deposit* greenish yellow. *Habitat* on soil, often near roadsides, but rare. *Edibility* unknown.

*CLATHRUS RUBER*

**CAGE FUNGUS**

GROUP Stinkhorns

FAMILY Clathraceae

SEASON EDIBILITY ?

SIMILAR SPECIES

The Devil's Fingers (*C. archeri*) has four radiating arms; edibility unknown.

### *PHALLUS IMPUDICUS*

Commonly located in woodlands by the nauseous odour; the spongy stalk emerges from a membranous egg, and bears at the apex a conical, slimy spore-mass which attracts flies.

*Egg* 1¼–1½ in/3–4 cm in diameter, ball-shaped, bearing root-like cords at the base, white, smooth and gelatinous, finally rupturing to release the stalk. *Stalk* rapidly expanding, up to 8 in/20 cm high, ⅝–1 in/1.5–2.5 cm in diameter, cylindrical, white, spongy and hollow. *Gleba* attached to stalk apex, more or less conical, with numerous branching ridges and covered initially by an olive-green-brown slime (strongly smelling of rotting meat), which is eaten by flies. *Habitat* among leaf-litter. *Edibility* the unopened egg is said to be edible.

*PHALLUS IMPUDICUS*

**COMMON STINKHORN**

GROUP Stinkhorns

FAMILY Phallaceae

SEASON EDIBILITY

SIMILAR SPECIES

Ravenel's Stinkhorn (*P. ravenelii*) has a pale yellowish stalk and a pinkish egg; inedible.

### *MUTINUS CANINUS*

A small delicate stinkhorn, with a yellow stalk and a red apex to the gleba.

*Egg* ⅜–¾ in/1–2 cm in diameter, ball-shaped, white, smooth, gelatinous, attached by root-like cords. *Stalk* 3½–4¾ in/9–12 cm tall when fully expanded, cylindrical, spongy, yellowish with a whitish base, hollow. *Gleba* acorn-shaped, covered by a dark olive-green slime, except for the red apex. *Spore deposit* pale yellowish. *Habitat* among woodland leaf-litter, usually in groups. *Edibility* inedible.

*MUTINUS CANINUS*

**DOG STINKHORN**

GROUP Stinkhorns

FAMILY Phallaceae

SEASON EDIBILITY

SIMILAR SPECIES

The Elegant Stinkhorn (*M. elegans*) has a slender, tapering gleba; inedible.

### *GYROMITRA ESCULENTA*

A conifer-wood species having a much-lobed, brain-like, reddish brown cap and a short, whitish stalk.

*Cap* 1¼–3½ in/3–9 cm across, much lobed and convoluted, brain-like in appearance, reddish brown. *Stem* ¾–2 × ¾–1½ in/2–5 × 2–4 cm, whitish or pale flesh coloured, often irregularly grooved, smooth or slightly scurfy. *Flesh* lacking a distinctive smell, whitish, thin, brittle, fruitbody internally chambered and hollow. *Habitat* with conifers, especially pines, growing on sandy soil. *Edibility* **poisonous; deadly if eaten raw and sometimes harmful even after cooking.**

*GYROMITRA ESCULENTA*

**TURBAN FUNGUS or FALSE MOREL**

GROUP Cup Fungi

FAMILY Saddle-cup (Helvellaceae)

SEASON EDIBILITY

SIMILAR SPECIES

There are several other species of *Gyromitra*, especially in America. These may differ in shape and colour, but require microscopic examination for positive identification.

## *HELVELLA CRISPA*

**A** common species having a whitish, saddle-shaped cap and deeply furrowed, whitish stalk.

*Cap* $\frac{3}{4}$–2 in/2–5 cm across, saddle-shaped, lobed, convoluted at the centre, lobes not attached to the stalk, fertile upper surface whitish, smooth, lower surface pale buff, minutely downy. *Stem* $\frac{3}{4}$–3 × $\frac{3}{8}$–1 in/2–8 × 1–2.5 cm, whitish, deeply furrowed with vertical grooves, finely downy above, hollow. *Flesh* thin, brittle, white. *Habitat* in deciduous or mixed woodland, often at path edges. *Edibility* edible but worthless.

*HELVELLA CRISPA*

**COMMON WHITE HELVELLA or FLUTED WHITE HELVELLA**

GROUP Cup Fungi

FAMILY Saddle-cup (Helvellaceae)

SEASON EDIBILITY

SIMILAR SPECIES

*H. lacunosa* is similar in form, but has a dark grey or blackish cap and stem, and lobes of the cap attached to the cap in places; inedible.

## *HELVELLA ACETABULUM*

**A** brownish, cup-shaped species with a whitish, furrowed stem, found in woodland in spring and early summer.

*Cup* $1\frac{1}{4}$–$2\frac{3}{4}$ in/3–7 cm across, $\frac{3}{8}$–$1\frac{1}{2}$ in/1–4 cm high, remaining deeply cup-shaped, inner fertile surface brown to chestnut-brown, smooth, outer surface slightly paler, finely downy. *Stem* $\frac{3}{8}$–$1\frac{1}{4}$ × $\frac{3}{8}$–$\frac{3}{4}$ in/1–3 × 1–2 cm, whitish, deeply furrowed, the ribs extending onto the lower surface of the cup, hollow. *Flesh* thin, brittle, pale brownish. *Habitat* woodland and heathland, especially on chalky soil. *Edibility* poisonous.

*HELVELLA ACETABULUM*

**RIBBED-STALK CUP**

GROUP Cup Fungi

FAMILY Saddle-cup (Helvellaceae)

SEASON EDIBILITY

SIMILAR SPECIES

*H. leucomelas* may be similar, but has a dark brown or blackish cup and much shorter stalk; poisonous.

## *HELVELLA ELASTICA*

**A** tall, slender species characterized by the pale, greyish yellow cap and non-furrowed, whitish stem.

*Cap* $\frac{3}{8}$–$1\frac{1}{4}$ in/1–3.5 cm across and high, irregularly saddle-shaped, 2 or 3 lobed, lobes deflexed, attached to the stem in places, upper surface pale grey or greyish yellow, drying dark brown, smooth, under surface paler, whitish, drying ochraceous, smooth. *Stem* $2\frac{1}{4}$–4 × $\frac{1}{8}$–$\frac{5}{16}$ in/6–10 × 0.3–0.8 cm, cylindrical, often slightly flattened, whitish, non-furrowed, hollow. *Flesh* thin, without a distinctive smell. *Habitat* in open woodland. *Edibility* inedible.

*HELVELLA ELASTICA*

**SMOOTH-STALKED HELVELLA**

GROUP Cup Fungi

FAMILY Saddle-cup (Helvellaceae)

SEASON EDIBILITY

SIMILAR SPECIES

*H. latispora* has a whitish or cream cap with margin entirely free from the stem; inedible.

## *VERPA CONICA*

**A** distinctive species, recognized by the brown, pendulous, thimble-shaped cap and whitish stem.

*Cap* 5⁄8–1½ in/1.5–4 cm high, 3⁄8–¾ in/1–2 cm wide, bell-shaped or thimble-shaped, dark brown or olive-brown, slightly wrinkled or furrowed, smooth, pendulous, attached only at the top of the stem. *Stem* 1¼–3 × ¼–3⁄8 in/3–8 × 0.6–1 cm, cylindrical or slightly thicker downwards, whitish or cream, with irregular bands of slightly darker granules, hollow. *Flesh* thin, brittle. *Habitat* among scrub, often with hawthorn and apple trees, and at roadsides, often on chalky soil. *Edibility* edible when cooked, but poor.

*VERPA CONICA*

**SMOOTH THIMBLE CAP**

GROUP Cup Fungi

FAMILY Morel (Morchellaceae)

SEASON EDIBILITY

SIMILAR SPECIES

*V. bohemica* is similar in form, but has a much more wrinkled cap; inedible.

## *DISCIOTIS VENOSA*

**A** large, brown, thick-fleshed, saucer-shaped species, with a strongly veined upper surface, and very short, stout stem.

*Fruitbody* up to 7 in/18 cm across, at first cup-shaped, becoming expanded and saucer-shaped, margin often reflexed and lobed, upper surface dark brown, with well-developed ribs and furrows radiating from the centre. *Outer surface* whitish, slightly scurfy. *Stem* very short, stout, often sunk in the soil and indistinct. *Flesh* thick, brittle, pale brownish, with a distinctive, unpleasant smell. *Habitat* on soil in woods and in grass. *Edibility* poisonous.

*DISCIOTIS VENOSA*

**VEINED CUP**

GROUP Cup Fungi

FAMILY Morel (Morchellaceae)

SEASON EDIBILITY

SIMILAR SPECIES

Species of *Peziza* may be similar in form, but mostly lack a stem, and have thinner flesh or are paler in colour. *Discina perlata* is usually smaller, with a less furrowed inner surface, and pale ochraceous outer surface; inedible.

## *OTIDEA ONOTICA*

**A**n ear-shaped species distinguished by its ochraceous, pink-tinged fruitbodies.

*Cup* 1½–4 in/4–10 cm high, 1¼–2¼ in/3–6 cm wide, elongated on one side and irregularly rabbit-ear shaped with short, indistinct stalk. *Fertile inner surface* smooth, ochraceous, distinctly tinged with pink, drying with a more pronounced pink tint. *Outer surface* similarly coloured, drying ochraceous, finely scurfy. *Flesh* thin, whitish, with no distinctive smell. *Habitat* in leaf-litter in deciduous or mixed woodland, often in clusters. *Edibility* inedible.

*OTIDEA ONOTICA*

**LEMON PEEL FUNGUS**

GROUP Cup Fungi

FAMILY Large Cup Fungi (Pezizaceae)

SEASON EDIBILITY

SIMILAR SPECIES

Species of *Flavoscypha* may be similar in form, but are brighter yellow and lack pink tints; inedible.

## FLAVOSCYPHA CANTHARELLA

**A**n ear-shaped species with a bright sulphur-yellow colours and whitish base.

*Cup* ¾–2 in/2–5 cm tall, ⅜–1¼ in/1–3 cm wide, greatly elongated on one side, rabbit-ear shaped, split to the base on the short side, with short, indistinct stalk. *Fertile inner surface* bright and sulphur-yellow, smooth. *Outer surface* similarly coloured, finely scurfy. *Stem* short, often indistinct, whitish-downy. *Flesh* thin, whitish, without a distinctive smell. *Habitat* on soil or in leaf-litter in woods and parks, sometimes in clusters. *Edibility* inedible.

*FLAVOSCYPHA CANTHARELLA*

**YELLOW HARE'S EAR**

GROUP Cup Fungi

FAMILY Large Cup Fungi (Pezizaceae)

SEASON EDIBILITY

SIMILAR SPECIES

*F. concinna* is similar in colour, but not so elongated on one side; inedible.

## PEZIZA BADIA

**A** cup- or saucer-shaped species having fruitbodies with dark olive-brown inner surfaces and red-brown outer surfaces.

*Fruitbody* 1¼–4 in/3–10 cm across, irregularly cup-shaped, often expanded and lobed, with margin undulating, lacking a stalk. *Inner surface* smooth, dark olive-brown. *Outer surface* dark brown or reddish brown, scurfy towards the margin. *Flesh* thin, reddish brown, brittle, lacking a distinctive smell. *Habitat* on bare, sandy soil, especially in woodlands, often growing in clusters. *Edibility* must be well cooked; poisonous when eaten raw.

*PEZIZA BADIA*

**PIG'S EARS**

GROUP Cup Fungi

FAMILY Large Cup Fungi (Pezizaceae)

SEASON EDIBILITY

SIMILAR SPECIES

*P. badioconfusa* and other brown species of *Peziza* are similar and difficult to distinguish without microscopic examination; inedible.

## SARCOSPHAERA CRASSA

**A** distinctive species recognized by the partially sunken, cup-shaped fruitbodies with violaceous inner surface.

*Fruitbody* 3–7 in/8–18 cm across, developing below the ground surface, lacking a stalk, at first closed, like a hollow ball, whitish on the surface, at maturity reaching the surface and splitting above into 6–8 star-like lobes. *Inner surface* violaceous, smooth. *Outer surface* whitish or cream coloured, finely downy, with attached fragments of soil. *Habitat* in woods on chalky soil. *Edibility* inedible.

*SARCOSPHAERA CRASSA*

**PINK CROWN**

GROUP Cup Fungi

FAMILY Large Cup Fungi (Pezizaceae)

SEASON EDIBILITY

SIMILAR SPECIES

Not likely to be confused with any other species.

## *TARZETTA CUPULARIS*

A cup-shaped, short-stalked, greyish buff species growing on damp soil in woodland.

*Fruitbody* 3/8–3/4 in/1–2 cm across, remaining deeply cup-shaped, short stalked, margin with small, irregular teeth. *Inner surface* pale buff, with greyish tinge, smooth. *Outer surface* similarly coloured but paler, finely but distinctly scurfy. *Stem* short, cylindrical, partly buried. *Flesh* thin, brittle. *Habitat* on damp soil, in woods or at path edges, solitary or gregarious. *Edibility* inedible.

*TARZETTA CUPULARIS*

**ELF CUP**

**GROUP** Cup Fungi

**FAMILY** Large Cup Fungi (Pezizaceae)

**SEASON** **EDIBILITY**

**SIMILAR SPECIES**

*T. catinus* is larger, lacks greyish tints and has a shorter stalk; inedible.

## *GEOPORA SUMNERIANA*

A distinctive species characterized by large, partially sunken, deeply cup-shaped fruitbodies which open by lobes.

*Fruitbody* 1½–3 in/4–8 cm across, developing below the soil surface as a closed, hollow ball, reaching the surface at maturity and splitting open irregularly into lobes which fold back to form an irregular star shape. *Inner surface* cream or pale yellowish, smooth. *Outer surface* pale brown, densely clothed with long, dark brown, matted hairs. *Flesh* thick, whitish, brittle. *Habitat* usually under cedars. *Edibility* inedible.

*GEOPORA SUMNERIANA*

**SUMNER'S EARTHCUP**

**GROUP** Cup Fungi

**FAMILY** Large Cup Fungi (Pezizaceae)

**SEASON** **EDIBILITY**

**SIMILAR SPECIES**

*G. foliacea* is usually smaller, and occurs in heavy soil in woodland; inedible.

CHAPTER THREE

# On The Ground In Grassland Or Open Spaces

## *CAMAROPHYLLUS NIVEUS*

**A** pure white wax-gill, recognized by the thick, decurrent gills.

*Cap* ¾–1½ in/2–4 cm in diameter, convex or flattened, pure white, smooth, with a striated margin. *Gills* decurrent, pure white, thick, well spaced and sometimes interveined. *Stem* ¾–1½ × 3/16–⅜ in/2–4 × 0.5–1 cm, cylindrical, white, smooth. *Flesh* white, soft, thin. *Spore deposit* white. *Habitat* sometimes found in large numbers in pastures and meadowland. *Edibility* edible but care should be taken to avoid confusion with the poisonous white *Clitocybe* species which have thin, crowded gills.

*CAMAROPHYLLUS NIVEUS*

**SNOWY MEADOW CAP**

**GROUP** Mushrooms and Toadstools

**FAMILY** Wax-gill (Hygrophoraceae)

**SEASON** **EDIBILITY**

**SIMILAR SPECIES**

*C. virgineus* is similar but more robust, while *C. russocoriaceus* has a strong, fragrant odour; both edible.

## *CAMAROPHYLLUS PRATENSIS*

**A** fleshy, orange-buff wax cap, which is regarded as a good, edible species.

*Cap* 1¼–3½ in/3–9 cm in diameter, convex, often with a raised centre, developing a wavy margin, orange-buff to tawny yellow, smooth, dry, sometimes cracking. *Gills* deeply decurrent, yellowish buff, thick and waxy, broadly spaced. *Stem* 2–3 × 3/16–⅜ in/5–8 × 0.5–1 cm, cylindrical or tapering at the base, similarly coloured to cap but often paler, smooth. *Flesh* thick, white, firm. *Spore deposit* white. *Habitat* in open grassland, sometimes also found in frondose woodland. *Edibility* excellent.

*CAMAROPHYLLUS PRATENSIS*

**BUFF MEADOW CAP or SALMON WAX CAP**

**GROUP** Mushrooms and Toadstools

**FAMILY** Wax-gill (Hygrophoraceae)

**SEASON** **EDIBILITY**

**SIMILAR SPECIES**

*C. colemannianus* is smaller, pinkish grey-brown and has a striated cap margin; edible.

## *HYGROCYBE CONICA*

**T**he most common of the brightly coloured wax caps which blacken at maturity.

*Cap* ¾–2 in/2–5 cm in diameter, conical and pointed, bright orange but progressively blackening with age, finely streaky, shiny, sticky. *Gills* adnexed, at first white later becoming greyish orange, and finally black, broad, widely spaced. *Stem* 2–4 × 3/16–⅜ in/5–10 × 0.5–1 cm, cylindrical, fibrous and splitting easily, yellowish. *Flesh* thin, watery, pale. *Spore deposit* white. *Habitat* commonly found anywhere among grass. *Edibility* inedible, to be avoided.

*HYGROCYBE CONICA*

**CONICAL WAX CAP or WITCH'S HAT**

**GROUP** Mushrooms and Toadstools

**FAMILY** Wax-gill (Hygrophoraceae)

**SEASON** **EDIBILITY**

**SIMILAR SPECIES**

The Blackening Meadow Cap (*H. nigrescens*) is more robust; poisonous. *H. conicoides* has a more rounded cap; inedible.

## *HYGROCYBE PSITTACINA*

**A** fairly common wax cap which may be difficult to recognize as it changes in colour and loses all greenish tints with age.

*Cap* ¾–2 in/2–5 cm in diameter, conical or bell-shaped, at first covered with a dark greenish slime, often with ochre, tawny red or pinkish tints. *Gills* adnate to short decurrent, greenish to yellowish, waxy, more or less spaced. *Stem* ¾–2 × 1/16–⅜ in/2–5 × 0.2–1 cm, cylindrical, greenish then developing yellow to orange or pink tints, slimy. *Flesh* thin, watery, pale. *Spore deposit* white. *Habitat* in coniferous or mixed woods, or grassland. *Edibility* edible.

*HYGROCYBE PSITTACINA*

**PARROT TOADSTOOL**

**GROUP** Mushrooms and Toadstools

**FAMILY** Wax-gill (Hygrophoraceae)

**SEASON** **EDIBILITY**

**SIMILAR SPECIES**

*H. citrinovirens* has a sulphur-yellow cap which acquires green tints; edibility unknown.

## *HYGROCYBE COCCINEA*

**A** grassland species, distinguished by the bright red cap, gills and stem.

*Cap* 1¼–2 in/3–5 cm in diameter, conical then expanding, blood-red to pinkish orange, smooth. *Gills* adnate, red to yellowish orange, thick, broad, moderately crowded. *Stem* 1¼–2¾ × 3/16–⅜ in/3–7 × 0.5–1 cm, cylindrical, scarlet-red, paler at the base, hollow, smooth. *Flesh* thin, yellow, watery. *Spore deposit* white. *Habitat* in grassland. *Edibility* edible.

*HYGROCYBE COCCINEA*

**SCARLET HOOD**

**GROUP** Mushrooms and Toadstools

**FAMILY** Wax-gill (Hygrophoraceae)

**SEASON** **EDIBILITY**

**SIMILAR SPECIES**

*H. punicea* is similarly coloured but more robust, with a conspicuously fibrous stem; edible.

## *CALOCYBE CARNEA*

**An** attractive species growing on lawns, distinguished by the contrasting pink cap and white gills.

*Cap* ¾–1½ in/2–4 cm in diameter, convex or flattened with a raised centre, rose-pink to flesh-pink, sometimes brownish, smooth and dry. *Gills* adnate to subdecurrent, white, narrow, crowded. *Stem* ¾–1¼ × ⅛ in/2–3 × 0.2–0.3 cm, slender, fibrous, pale pinkish. *Flesh* thick, white. *Spore deposit* white. *Habitat* on lawns or in meadows. *Edibility* worthless.

*CALOCYBE CARNEA*

**PINK CALOCYBE**

**GROUP** Mushrooms and Toadstools

**FAMILY** Tricholoma (Tricholomataceae)

**SEASON** **EDIBILITY**

**SIMILAR SPECIES**

*C. persicolor* is more fleshy, often grows in tufts and has a hairy stem base; inedible.

## CALOCYBE GAMBOSA

This large, whitish mushroom is usually much sought after as it is one of the few, good edible species to occur in spring.

*Cap* 2–4 in/5–10 cm in diameter, convex with a wavy, inrolled margin, white to pale brown, smooth and dry. *Gills* sinuate, white to cream, narrow and densely crowded. *Stem* 1¼–3 × ⅜–1 in/3–8 × 1–2.5 cm, short and stocky, white, smooth, solid. *Flesh* thick, white, with a strong smell of damp flour. *Spore deposit* white. *Habitat* among grass, preferring chalky soil. *Edibility* good.

*CALOCYBE GAMBOSA*

**ST GEORGE'S MUSHROOM**

**GROUP** Mushrooms and Toadstools

**FAMILY** Tricholoma (Tricholomataceae)

**SEASON** **EDIBILITY**

**SIMILAR SPECIES**

The springtime appearance is likely to overcome confusion with any similar species, but avoid collecting at the same time as the reddening *Inocybe* species and the white *Clitocybe* species, which grow in similar situations; poisonous.

## LYOPHYLLUM DECASTES

A grey-brown species, forming dense clusters in grassland and on bare soil.

*Cap* 1¼–4¾ in/3–12 cm in diameter, convex soon flattening, greyish to reddish brown, smooth, with an incurved margin becoming upturned and wavy. *Gills* adnate to short decurrent, whitish or yellowish, broad and crowded. *Stem* 2–4 × 3⁄16–¾ in/5–10 × 0.5–2 cm, short and stocky, whitish, smooth, solid. *Flesh* thick, pale, fibrous. *Spore deposit* white. *Habitat* forming clusters, often near buried roots or stumps. *Edibility* said to be edible.

*LYOPHYLLUM DECASTES*

**CLUSTERED GREY GILL or FRIED CHICKEN MUSHROOM**

**GROUP** Mushrooms and Toadstools

**FAMILY** Tricholoma (Tricholomataceae)

**SEASON** **EDIBILITY**

**SIMILAR SPECIES**

*L. fumosum* is very similar, with grey gills, while *L. loricatum* has a dark, almost black cap; edibility unknown for both species.

## CLITOCYBE RIVULOSA

Forms fairy rings on lawns and may be found mixed with the Fairy Ring Champignon *Marasmius oreades,* (page 76).

*Cap* ¾–2 in/2–5 cm in diameter, flat with a depressed or raised centre, white becoming greyish pink or yellowish, developing indistinct wrinkles in concentric zones. *Gills* short decurrent, whitish, very crowded. *Stem* ¾–2 × 3⁄16–⅜ in/2–5 × 0.5–1 cm, short cylindrical, similarly coloured to cap. *Flesh* white, soft. *Spore deposit* white. *Habitat* in troops in short grass. *Edibility* poisonous, causing muscarine poisoning.

*CLITOCYBE RIVULOSA*

**CRACKING CLITOCYBE**

**GROUP** Mushrooms and Toadstools

**FAMILY** Tricholoma (Tricholomataceae)

**SEASON** **EDIBILITY**

**SIMILAR SPECIES**

The poisonous Ivory Clitocybe (*C. dealbata*) is very closely related but is white with a more distinctly zoned cap. Avoid confusing either with the edible *Marasmius oreades*.

## *OMPHALINA ERICETORUM*

**A**lthough common, the small *Omphalina* species are easily overlooked. They are characterized by their small size and by their pale decurrent gills.

*Cap* ⅜–1¼ in/1–3 cm in diameter, flat to depressed to almost funnel-shaped, brownish becoming yellowish, radially furrowed, moist, with a wavy margin. *Gills* decurrent, triangular, cream-coloured, spaced. *Stem* ⅜–1¼ × ⅛ in/1–3 × 0.2–0.3 cm, cylindrical, yellowish to brown, smooth, with a hairy base. *Flesh* thin, white. *Habitat* among moss, on peaty soil, and always associated with the lichen *Botrydina*. *Edibility* inedible.

*OMPHALINA ERICETORUM*

**UMBRELLA NAVEL CAP**

**GROUP** Mushrooms and Toadstools

**FAMILY** Tricholoma (Tricholomataceae)

**SEASON** **EDIBILITY**

**SIMILAR SPECIES**

*O. rustica* has narrow, grey gills and the cap is greyish brown, while *O. pyxidata* has an orange-brown cap with dark striae; both inedible.

## *RICKENELLA FIBULA*

**T**his minute species is easily overlooked but is often to be found in large numbers among moss.

*Cap* ⅛–⅜ in/0.3–1 cm in diameter, convex then depressed in the centre, radially striated, bright yellowish orange, dry. *Gills* deeply decurrent, pale orange, narrow, well spaced. *Stem* ⅜–1½ × 1/16 in/1–4 × 0.1–0.2 cm, thread-like, yellowish orange, hollow and fragile, smooth. *Flesh* membranous, pale. *Spore deposit* white. *Habitat* very common, growing in troops among moss. *Edibility* edible but worthless.

*RICKENELLA FIBULA*

**ORANGE MOSS AGARIC or CARPET-PIN MYCENA**

**GROUP** Mushrooms and Toadstools

**FAMILY** Tricholoma (Tricholomataceae)

**SEASON** **EDIBILITY**

**SIMILAR SPECIES**

The variety *swartzii* has grey and black tints, especially at the cap centre and stem apex; inedible.

## *MARASMIUS OREADES*

**A**n unpopular species with gardeners owing to its habit of forming extensive fairy rings which last for many years.

*Cap* ¾–2 in/2–5 cm in diameter, bell-shaped, finally expanding with a wavy margin, pale rusty brown but drying much paler, dry, smooth. *Gills* adnexed, white, broad, widely spaced. *Stem* ¾–2¾ × 1/16–3/16 in/2–7 × 0.2–0.4 cm, thin, tough, similarly coloured to cap, dry. *Flesh* thin, white, with a smell recalling hay. *Spore deposit* white. *Habitat* growing in large numbers in grassland, especially on lawns where it frequently forms extensive, perennial fairy rings. *Edibility* edible and good, often used in quantity in stews, casseroles, and such like.

*MARASMIUS OREADES*

**FAIRY RING CHAMPIGNON**

**GROUP** Mushrooms and Toadstools

**FAMILY** Tricholoma (Tricholomataceae)

**SEASON** **EDIBILITY**

**SIMILAR SPECIES**

A distinctive species but sometimes growing intermixed with the poisonous, white *Clitocybe* species which have decurrent, crowded gills. Care must be taken when collecting this species.

## *MYCENA LEPTOCEPHALA*

**A** very common *Mycena* species, with an alkaline smell and grey colours.

*Cap* ¾–1¼ in/2–3 cm in diameter, bell-shaped, greyish brown, paler at the margin, radially striated. *Gills* adnate, greyish, moderately spaced. *Stem* 2–2¾ × 1/16 in/5–7 × 0.1–0.2 cm, slender, cylindrical, hollow and brittle, similarly coloured to cap or paler. *Flesh* thin, grey, watery, brittle, with a strong alkaline smell. *Spore deposit* white. *Habitat* solitary on soil among short grass, and often pine-needles. *Edibility* inedible.

*MYCENA LEPTOCEPHALA*

**NITROUS MYCENA**

**GROUP** Mushrooms and Toadstools

**FAMILY** Tricholoma (Tricholomataceae)

**SEASON** **EDIBILITY**

**SIMILAR SPECIES**

The Stump Mycena (*M. alcalina*) is similar but forms small tufts on rotting woods; inedible.

## *LEUCOPAXILLUS GIGANTEUS*

**A** very large and stout, short-stemmed white mushroom, with crowded gills descending the stem; often forming large fairy rings.

*Cap* 4–16 in/10–40 cm in diameter, flattened to funnel-shaped, white or nearly so, at first moist, smooth, with a thin, inrolled margin. *Gills* decurrent, white to pale buff, drying brownish, densely crowded. *Stem* 1¼–4 × 1¼–2 in/3–10 × 3–5 cm, short and thick, solid, whitish and smooth. *Flesh* thick, firm to hard. *Spore deposit* white. *Habitat* in open woodland. *Edibility* edible but only after first boiling; some persons show an allergic reaction.

*LEUCOPAXILLUS GIGANTEUS*

**GIANT CLITOCYBE**

**GROUP** Mushrooms and Toadstools

**FAMILY** Tricholoma (Tricholomataceae)

**SEASON** **EDIBILITY**

**SIMILAR SPECIES**

This species could be confused with the Fleecy Milk-cap (*Lactarius vellereus*, page 49) or the Peppery Milk-cap (*L. piperatus*), *both of which release an acrid, white latex and are inedible.*

## *ENTOLOMA RHODOPOLIUM*

**A** tall but fragile mushroom, distinguished by the pinkish gills and an odour of meal.

*Cap* ¾–4 in/2–10 cm in diameter, convex with a raised or depressed centre, greyish brown with yellowish tints, smooth and silky. *Gills* adnate, at first white becoming pink as the spores mature, crowded. *Stem* 2–4 × 3/16–3/8 in/5–10 × 0.5–1 cm, cylindrical, white, smooth. *Flesh* thick and brittle, with a strong smell recalling damp flour. *Spore deposit* pink. *Habitat* small groups among leaf-litter in deciduous woods. *Edibility* poisonous, causing stomach upsets. All pink-gilled species are best avoided.

*ENTOLOMA RHODOPOLIUM*

**ROSY ENTOLOMA**

**GROUP** Mushrooms and Toadstools

**FAMILY** Entoloma (Entolomataceae)

**SEASON** **EDIBILITY**

**SIMILAR SPECIES**

The Strong-scented Entoloma (*E. nidorosum*) is paler, less robust, and has a more alkaline smell; poisonous.

## *NOLANEA SERICEA*

**A** pink-spored toadstool, commonly found on lawns throughout the autumn months.

*Cap* $\frac{3}{4}$–$1\frac{1}{2}$ in/2–4 cm in diameter, convex then expanded with a wavy margin, dark brown and finely striate when moist but drying much paler and becoming shiny-silky. *Gills* adnexed, pale grey becoming progressively pink, moderately crowded. *Stem* $\frac{3}{4}$–2 × $\frac{1}{8}$ in/2–5 × 0.2–0.3 cm, slender, fibrous, greyish brown. *Flesh* thin, greyish, watery, with a strong smell of meal. *Spore deposit* pink. *Habitat* common, scattered among short grass and frequently encountered on lawns. *Edibility* inedible.

| *NOLANEA SERICEA* |
| --- |
| **SILKY NOLANEA** |
| GROUP Mushrooms and Toadstools |
| FAMILY Entoloma (Entolomataceae) |
| SEASON / EDIBILITY |
| SIMILAR SPECIES<br>There are a number of slender *Nolanea* species growing in grassland, but most are less common and require detailed examination under the microscope for identification. |

## *LEPTONIA SERRULATA*

**T**he most common of the blue *Leptonia* species, with a black, saw-like gill-edge.

*Cap* $\frac{3}{8}$–$1\frac{1}{4}$ in/1–3.5 cm in diameter, convex then expanding to flat with a wavy margin, dark bluish black, with small scales especially at the centre, dry. *Gills* adnate, pale blue then greyish pink with a black edge. *Stem* $\frac{3}{4}$–2 × $\frac{1}{16}$–$\frac{3}{16}$ in/2–5 × 0.2–0.4 cm, slender, paler than cap, with dark bluish grey fibrils, and a white, hairy base. *Flesh* pale, brittle. *Spore deposit* pink. *Habitat* fairly common in grassland. *Edibility* inedible, with an unpleasant taste.

| *LEPTONIA SERRULATA* |
| --- |
| **SAW GILLED LEPTONIA or BLUE-TOOTHED LEPTONIA** |
| GROUP Mushrooms and Toadstools |
| FAMILY Entoloma (Entolomataceae) |
| SEASON / EDIBILITY |
| SIMILAR SPECIES<br>*L. euchroa* is bright violaceous and grows on dead wood, while *L. lampropus* has a grey-brown cap; both inedible. |

## *LEUCOAGARICUS NAUCINUS*

**A** fleshy grassland fungus, differing from the true mushrooms in having gills which remain white.

*Cap* 2–4 in/5–10 cm in diameter, broadly convex to flattened, dry, pure white and smooth. *Gills* free, white, broad, crowded. *Stem* 2–3 × $\frac{3}{16}$–$\frac{3}{8}$ in/5–8 × 0.5–1 cm, short and stocky, smooth, white, bearing a white, membranous ring on the upper region. *Flesh* thick, white, firm. *Spore deposit* white. *Habitat* among grass, especially on lawns. *Edibility* poisonous, causing stomach upsets.

| *LEUCOAGARICUS NAUCINUS* |
| --- |
| **SMOOTH LEPIOTA** |
| GROUP Mushrooms and Toadstools |
| FAMILY Mushroom and Lepiota (Agaricaceae) |
| SEASON / EDIBILITY |
| SIMILAR SPECIES<br>This species is easily confused with the true mushrooms (*Agaricus* species), which differ in having gills which darken. Check first with a spore print before attempting to eat. |

## LEUCOCOPRINUS BREBISSONII

**A** tall, delicate toadstool, with a white cap bearing a dark central spot.

*Cap* ¾–2 in/2–5 cm in diameter, conical then flattened with a slightly raised centre, pure white with a very dark greyish brown disc at the centre and scattered minute brown scales, and the margin strongly striate halfway to the centre. *Gills* free, white, narrow, more or less crowded. *Stem* 1½–3 × 1/16–3/16 in/4–8 × 0.2–0.4 cm, cylindrical, white, hollow, smooth, bearing a small, white, membranous ring in the middle zone. *Flesh* membranous, white. *Spore deposit* white. *Habitat* on bare soil and among grass, often found by the side of paths but not common. *Edibility* worthless.

*LEUCOCOPRINUS BREBISSONII*

### OX-EYE LEPIOTA

**GROUP** Mushrooms and Toadstools

**FAMILY** Mushroom and Lepiota (Agaricaceae)

**SEASON** **EDIBILITY**

**SIMILAR SPECIES**

This species does not have a swollen stem base as in related *Leucocoprinus* species, and the poisonous Stinking Parasol (*Lepiota cristata*, page 35) has a smaller stature and lacks the striate margin.

## LEUCOCOPRINUS CEPAESTIPES

**A** white, tufted mushroom, it is recognized by the swollen stem base. Found during the summer months in warmer regions, it often also occurs in hot-houses all year round.

*Cap* 1¼–2¾ in/3–7 cm in diameter, bell-shaped but finally expanding with a raised centre and a wavy margin, white, radially ridged, but covered with small, powdery scales. *Gills* free, whitish, crowded. *Stem* 2–4 × 1/8–3/16 in/5–10 × 0.3–0.5 cm, slender but with a swollen base (up to 1.5 cm across), white bruising yellowish brown, somewhat powdery, and bearing a small, thin ring on the upper region. *Flesh* thin, white, soft. *Spore deposit* white. *Habitat* found among grass or leaf-litter. *Edibility* inedible and possibly poisonous.

*LEUCOCOPRINUS CEPAESTIPES*

### ONION-STALKED LEPIOTA

**GROUP** Mushrooms and Toadstools

**FAMILY** Mushroom and Lepiota (Agaricaceae)

**SEASON** **EDIBILITY**

**SIMILAR SPECIES**

*L. brebissonii* (see this page) is smaller, has a black spot in the centre of the cap and a non-swollen stem base; inedible.

## LEUCOCOPRINUS LUTEUS

**The** fruitbody is uniformly sulphur-yellow, with a powdery cap, and grows in tufts. The fungus often occurs in hot-houses all year round and in warmer regions during the summer months.

*Cap* 1¼–2 in/3–5 cm in diameter, bell-shaped or conical, soft and delicate, sulphur-yellow, with powdery, loose scales and a grooved margin. *Gills* free, slightly paler yellow, thin and delicate, moderately crowded. *Stem* 2–4 × 1/8–3/16 in/5–10 × 0.3–0.5 cm, slender except for the swollen base, sulphur-yellow, smooth, with a small, membranous ring attached to the upper region. *Flesh* thin, soft, yellowish. *Spore deposit* white. *Habitat* among grass, on bare soil or among leaf-litter. *Edibility* inedible, possibly poisonous.

*LEUCOCOPRINUS LUTEUS*

### YELLOW COTTONY AGARIC

**GROUP** Mushrooms and Toadstools

**FAMILY** Mushroom and Lepiota (Agaricaceae)

**SEASON** **EDIBILITY**

**SIMILAR SPECIES**

Also known under the name *L. birnbaumii*, and very closely related to the white *L. cepaestipes* (see this page), the two species often grow together; inedible.

## *MACROLEPIOTA EXCORIATA*

**A** large, fleshy mushroom with white gills, sometimes found growing in large troops in meadows.

*Cap* 1½–3 in/4–8 cm in diameter, conical to convex, usually with a raised centre, whitish to cream or with greyish tints, developing very fine scales. *Gills* free, white, broad and crowded. *Stem* 2¾–3 × ⅜–⅝ in/7–8 × 1–1.5 cm, cylindrical, white, smooth and silky, bearing a large, membranous, double-ring. *Flesh* thick, soft and white. *Spore deposit* white. *Habitat* among grass, often on cultivated land. *Edibility* edible and good.

*MACROLEPIOTA EXCORIATA*

**FLAKY LEPIOTA**

**GROUP** Mushrooms and Toadstools

**FAMILY** Mushroom and Lepiota (Agaricaceae)

**SEASON** **EDIBILITY**

**SIMILAR SPECIES**

Tends to have a shorter stem and a more stocky aspect than other related *Macrolepiota* species.

## *MACROLEPIOTA GRACILENTA*

**O**ne of the tall *Macrolepiota* species, characterized by the small, tawny brown scales.

*Cap* 2¾–6 in/7–15 cm in diameter, convex with a raised centre, at first pale brown and smooth but the surface soon breaking up into small scales towards the margin and exposing a whitish background. *Gills* free, white, broad, and densely crowded. *Stem* 4–6 × ⅜–⅝ in/10–15 × 1–1.5 cm, tall, cylindrical with a swollen base, covered with tiny, brown scales, and bearing a large, membranous, movable ring. *Flesh* thick, white, soft. *Spore deposit* white. *Habitat* forming troops on pastureland. *Edibility* edible and good.

*MACROLEPIOTA GRACILENTA*

**SLENDER LEPIOTA**

**GROUP** Mushrooms and Toadstools

**FAMILY** Mushroom and Lepiota (Agaricaceae)

**SEASON** **EDIBILITY**

**SIMILAR SPECIES**

The Bossed Lepiota (*M. mastoidea*) is shorter, and the cap is coloured yellowish cream especially towards the margin; edible.

## *CHLOROPHYLLUM MOLYBDITES*

**A** large, white, fleshy mushroom found on lawns. It is frequently confused with the parasol mushrooms, but differs in the green gills which are a unique feature.

*Cap* 2–6 (–8) in/5–15 (–20) cm in diameter, at first ball-shaped soon expanding to broadly convex, initially covered by a dark brown layer which breaks up to expose a pure white background, leaving thick dark concentric scales at the centre. *Gills* free, pale buff but later becoming deep green, staining reddish when bruised, broad, very crowded. *Stem* 4–8 × ⅝–1 in/10–20 × 1.5–2.5 cm, easily detached from the cap, cylindrical except for the swollen base, whitish to pale greyish brown, fibrous, with a large, thick fleshy ring, attached to the upper region but becoming free and movable. *Flesh* pale pinkish, reddening on bruising, thick, with a pleasant odour. *Spore deposit* pea-green. *Habitat* scattered and sometimes in large numbers among grass, including lawns in warm localities; not present in Europe. *Edibility* poisonous, can cause serious stomach upsets and vomiting.

*CHLOROPHYLLUM MOLYBDITES*

**GREEN-SPORED LEPIOTA**

**GROUP** Mushrooms and Toadstools

**FAMILY** Mushroom and Lepiota (Agaricaceae)

**SEASON** **EDIBILITY**

**SIMILAR SPECIES**

This species is frequently confused with the edible true parasol mushroom (*Macrolepiota* species), which have white instead of green gills. Always take a spore print to check for identity.

## *AGARICUS ARVENSIS*

**A** common, grassland true mushroom, which discolours yellowish on bruising.

*Cap* 2¾–6 in/7–15 cm in diameter, convex, finally expanding, white, bruising pale yellow when the surface is rubbed, smooth to slightly scaly, with remnants of veil hanging from the margin. *Gills* free, greyish darkening to chocolate-brown, broad and crowded. *Stem* 2¾–4¾ × ⅜–¾ in/7–12 × 1–2 cm, cylindrical, smooth, white, bruising yellow, bearing a white, membranous, thick ring on the upper region. *Flesh* thick, white, with a smell of bitter almonds. *Spore deposit* blackish brown. *Habitat* common and scattered in pastureland. *Edibility* edible and good, but care must be taken over species which bruise yellow.

*AGARICUS ARVENSIS*

### HORSE MUSHROOM

**GROUP** Mushrooms and Toadstools

**FAMILY** Mushroom and Lepiota (Agaricaceae)

**SEASON** **EDIBILITY**

**SIMILAR SPECIES**

The poisonous Yellow-staining Mushroom (*A. xanthodermus,* see page 82) has a large ring, and an unpleasant smell, and the stem base discolours bright yellow. The edible Field Mushroom (*A. campestris,* see this page) does not bruise yellow.

## *AGARICUS CAMPESTRIS*

**T**he best-known mushroom, commonly illustrated in biology textbooks, distinguished by the reddening flesh, and the small, simple ring on the stem.

*Cap* 1¼–3 in/3–8 cm in diameter, convex becoming almost flat, pure white or sometimes with greyish tints at the centre, dry, smooth or indistinctly scaly in old specimens. *Gills* free, bright pink becoming dark chocolate-brown, broad and crowded. *Stem* 1¼–2¼ × ⅜–⅝ in/3–6 × 1–1.5 cm, short and cylindrical or tapering at the base, white, smooth, bearing a small, thin ring which easily disappears on weathering. *Flesh* thick, white discolouring pinkish when broken open, with a pleasant smell. *Spore deposit* blackish brown. *Habitat* common, scattered in open grassland. *Edibility* good.

*AGARICUS CAMPESTRIS*

### FIELD MUSHROOM

**GROUP** Mushrooms and Toadstools

**FAMILY** Mushroom and Lepiota (Agaricaceae)

**SEASON** **EDIBILITY**

**SIMILAR SPECIES**

The Horse Mushroom (*A. arvensis,* see this page) occurs in similar situations but is larger and bruises yellow, and the Cultivated Mushroom (*A. bisporus*) has a much larger ring; both edible.

## *AGARICUS BITORQUIS*

**A** fleshy, white, true mushroom, often found in urban situations, and recognized by the presence of two rings on the stem.

*Cap* 2–6 in/5–15 cm in diameter, convex to flattened and finally depressed in the centre, white to pale greyish brown, dry, smooth or cracked. *Gills* free, pale pink darkening to blackish brown, broad, very crowded. *Stem* ¾–2 × ⅜–1¼ in/2–5 × 1–3 cm, short and stocky, white, smooth, with a double-ring which has a thick upper edge and a free lower edge. *Flesh* thick, white, firm. *Habitat* growing on firm ground, often at base of trees, and sometimes between paving stones. *Edibility* edible and good.

*AGARICUS BITORQUIS*

### PAVEMENT MUSHROOM and SPRING AGARICUS

**GROUP** Mushrooms and Toadstools

**FAMILY** Mushroom and Lepiota (Agaricaceae)

**SEASON** **EDIBILITY**

**SIMILAR SPECIES**

Other true mushrooms (*Agaricus* species) have a simple ring. Care should be taken not to confuse the double-ring in young specimens with the volva found in poisonous *Amanita* species.

## *AGARICUS XANTHODERMUS*

**A** large and common, true mushroom, which bruises yellow, and can be immediately recognized by the bright yellow discoloration of the flesh at the stem base.

*Cap* 2¼–4¾ in/6–12 cm in diameter, strongly convex then gradually expanding to flat, white or perhaps with a pale greyish centre, dry, bruising bright yellow when rubbed. *Gills* free, white becoming pinkish and finally dark chocolate-brown. *Stem* 2–4¾ × ⅜–¾ in/5–12 × 1–2 cm, cylindrical with a slightly swollen base, white, smooth, bruising yellow; bearing a very large, white, membranous ring on the upper region. *Flesh* white, soft, with the base of the stem immediately discolouring bright golden-yellow when broken open; with an unpleasant carbolic odour. *Spore deposit* purplish brown. *Habitat* often under hedges or beside paths, common. *Edibility* poisonous, causing severe stomach upsets and even coma in extreme cases.

*AGARICUS XANTHODERMUS*

**YELLOW-STAINING MUSHROOM or YELLOW-FOOT AGARICUS**

**GROUP** Mushrooms and Toadstools

**FAMILY** Mushroom and Lepiota (Agaricaceae)

**SEASON** **EDIBILITY**

**SIMILAR SPECIES**

The poisonous Flat-topped Agaricus (*A. placomyces*) is very closely related but has small, sooty brown scales over the cap surface.

## *COPRINUS ATRAMENTARIUS*

**A** large ink-cap, usually forming tufts at the base of tree-stumps.

*Cap* 2–2¾ in/5–7 cm in diameter, bell-shaped then expanding slightly, grey, with a few tiny scales at the centre, and finely grooved towards the margin. *Gills* free, at first white then progressively grey from the edge inwards and finally dissolving into a black, inky liquid, very densely crowded. *Stem* 2¾–6 × ⅜–⅝ in/7–15 × 1–1.5 cm, cylindrical, white, hollow, with a faint ring-zone at the base. *Flesh* thin, greyish, brittle. *Spore deposit* black. *Habitat* forming clusters at or near tree-stumps. *Flesh* thin, greyish, brittle. *Edibility* young specimens are edible and good, but alcohol must **never** be consumed with the meal.

*COPRINUS ATRAMENTARIUS*

**COMMON INK-CAP or ALCOHOL INK-CAP**

**GROUP** Mushrooms and Toadstools

**FAMILY** Ink-cap (Coprinaceae)

**SEASON** **EDIBILITY**

**SIMILAR SPECIES**

*C. acuminatus* is more slender and has a pointed cap; *C. insignia* has warted spores.

## *COPRINUS COMATUS*

**A** well-known, tufted species, often called the "Lawyer's Wig" owing to the tiers of curled, white scales on the cap.

*Cap* tall and cylindrical, up to 5½ in/14 cm high, 1¼–2 in/3–5 cm across, with a rounded apex, only expanding slightly as it matures, white, covered with curled, woolly scales, dry, blackening from the margin upwards as the gills deliquesce. *Gills* free, white becoming grey then black from the edge, and dissolving into a black ink, very densely crowded. *Stem* 2¼–8 × ⅜–¾ in/6–20 × 1–2 cm, cylindrical, white, hollow, bearing a movable ring towards the base. *Flesh* white, soft. *Spore deposit* black. *Habitat* commonly found in grassland, sometimes in very large numbers, and frequently between paving stones in towns. *Edibility* good when young.

*COPRINUS COMATUS*

**SHAGGY INK-CAP**

**GROUP** Mushrooms and Toadstools

**FAMILY** Ink-cap (Coprinaceae)

**SEASON** **EDIBILITY**

**SIMILAR SPECIES**

The Magpie Ink-cap (*C. picaceus*, page 40) grows solitary in beechwoods, and the cap scales form white patches on a black background; inedible.

## *COPRINUS PLICATILIS*

**A** small, delicate ink-cap, frequently found on lawns and recognized by the grooved cap and well-spaced gills.

*Cap* $\frac{3}{4}$–$1\frac{1}{4}$ in/2–3 cm in diameter, convex soon becoming flattened, with a sunken brown centre, otherwise grey and radially pleated. *Gills* free from the stem but joined together by a collar, narrow, dark grey, well spaced. *Stem* 2–$2\frac{3}{4}$ × $\frac{1}{8}$ in/5–7 × 0.2–0.3 cm, cylindrical, slender, fragile, hollow, white. *Flesh* membranous. *Spore deposit* black. *Habitat* usually solitary on lawns and grass verges. *Edibility* worthless.

*COPRINUS PLICATILIS*

**LITTLE JAP UMBRELLA**

**GROUP** Mushrooms and Toadstools

**FAMILY** Ink-cap (Coprinaceae)

**SEASON** **EDIBILITY**

**SIMILAR SPECIES**

The poisonous Brown Hay Cap (*Panaeolina foenisecii*, see this page) grows in similar situations but is more stocky and lacks grooves on the cap.

## *PANAEOLINA FOENISECII*

**A** very common toadstool found on many lawns from summer onwards; the dark mottled gills and zoned cap are distinctive.

*Cap* $\frac{3}{8}$–$\frac{3}{4}$ in/1–2 cm in diameter, convex to bell-shaped, dark greyish when moist drying to a pale clay-brown from the centre outwards and often appearing zoned, dry and smooth. *Gills* adnate, greyish to dark brown, mottled, moderately crowded. *Stem* $\frac{3}{4}$–2 × $\frac{1}{16}$–$\frac{3}{16}$ in/2–5 × 0.2–0.4 cm, short, cylindrical, paler than the cap, smooth. *Flesh* thin, whitish. *Spore deposit* blackish brown. *Habitat* ubiquitous on lawns. *Edibility* inedible, and possibly mildly hallucinogenic in children.

*PANAEOLINA FOENISECII*

**BROWN HAY CAP or MOWER'S MUSHROOM**

**GROUP** Mushrooms and Toadstools

**FAMILY** Ink-cap (Coprinaceae)

**SEASON** **EDIBILITY**

**SIMILAR SPECIES**

The genus is doubtfully distinct from *Panaeolus*, of which all species have dark, mottled gills, and generally a much longer stem.

## *LACRYMARIA VELUTINA*

**A** large, fleshy mushroom, growing in tufts, and recognized by the black cottony ring on the stem and the black droplets on the gill-edge.

*Cap* $1\frac{1}{2}$–4 in/4–10 cm in diameter, bell-shaped, expanding, often with a raised centre, clay-brown or orange-brown, radially streaky, with an inrolled woolly margin which is at first white but blackens as the spores are released. *Gills* adnate, brown to blackish, mottled, broad and crowded, with a white edge which bears black watery droplets. *Stem* $2\frac{1}{4}$–$4\frac{3}{4}$ × $\frac{3}{16}$–$\frac{3}{8}$ in/6–12 × 0.5–1 cm, cylindrical, pale brown, fibrous, bearing a cobweb-like veil forming a ring-zone on the upper region. *Flesh* thick, white. *Spore deposit* black. *Habitat* forming tufts in fields and open woodland, and often near buildings. *Edibility* said to be edible, but not recommended.

*LACRYMARIA VELUTINA*

**WEEPING WIDOW**

**GROUP** Mushrooms and Toadstools

**FAMILY** Ink-cap (Coprinaceae)

**SEASON** **EDIBILITY**

**SIMILAR SPECIES**

*L. polytricha* is more orange coloured and associated with spruce stumps; inedible.

## *AGROCYBE PRAECOX*

**A** common springtime mushroom, with brown gills and a ring on the stem.

*Cap* 1¼–4 in/3–10 cm in diameter, broadly convex then becoming flattened, with a wavy margin, pale reddish brown, dry, smooth. *Gills* adnexed, whitish to cinnamon-brown, broad and crowded. *Stem* 2–4 × 3/16–⅝ in/5–10 × 0.5–1.5 cm, cylindrical, whitish, smooth, bearing a conspicuous membranous ring on the upper region. *Flesh* fairly thick, off-white, with an odour of damp flour. *Spore deposit* cinnamon-brown. *Habitat* often in scattered groups among grass and on bare soil. *Edibility* not recommended.

*AGROCYBE PRAECOX*

**SPRING AGARIC**

**GROUP** Mushrooms and Toadstools

**FAMILY** Bolbitius (Bolbitiaceae)

**SEASON** **EDIBILITY**

**SIMILAR SPECIES**

The Hemispheric Agrocybe (*A. semiorbicularis*) is smaller, yellowish buff and lacks a ring on the stem; inedible.

## *BOLBITIUS VITELLINUS*

**A** delicate fungus with a fragile, shiny cap and stem, characterized by the bright yellow centre to the cap.

*Cap* ¾–1½ in/2–4 cm in diameter, conical soon expanded to flat, bright yellow, but becoming much paler with age, slimy, almost transparent at the grooved margin. *Gills* adnexed to free, cinnamon to cinnamon-brown, narrow, moderately spaced. *Stem* 1¼–2¾ × ⅛ in/3–7 × 0.2–0.4 cm, slender, cylindrical, hollow, pale yellow, smooth. *Flesh* thin, membranous. *Spore deposit* rusty brown. *Habitat* often in open parkland and associated with all kinds of dung, including bird droppings. *Edibility* edible but worthless.

*BOLBITIUS VITELLINUS*

**YELLOW COW-PAT TOADSTOOL**

**GROUP** Mushrooms and Toadstools

**FAMILY** Bolbitius (Bolbitiaceae)

**SEASON** **EDIBILITY**

**SIMILAR SPECIES**

This delicate species tends to develop gills which deliquesce in wet weather, but these always remain brown, not black as in the ink-caps.

## *CONOCYBE TENERA*

**The** conical, pale brown cap, brown gills, and the grassland habitat identify this species.

*Cap* ⅜–1¼ in/1–3 cm in diameter, conical and remaining so, ochre-brown, sometimes paler, smooth and dry. *Gills* adnate, cinnamon-brown, narrow, fairly crowded. *Stem* 1½–3 × ⅛ in/4–8 × 0.2–0.3 cm, slender, cylindrical, pale brown, darker towards the base. *Flesh* thin, cream-coloured. *Spore deposit* rusty brown. *Habitat* scattered, among grass or along path edges in woodland. *Edibility* inedible.

*CONOCYBE TENERA*

**BROWN CONE CAP**

**GROUP** Mushrooms and Toadstools

**FAMILY** Bolbitius (Bolbitiaceae)

**SEASON** **EDIBILITY**

**SIMILAR SPECIES**

Many cone caps are similar in appearance and require careful examination under a microscope to distinguish between them.

## *CONOCYBE FILARIS*

**A** small, slender cone cap, characterized by a membranous ring on the stem.

*Cap* $\frac{3}{16}$–$\frac{3}{8}$ in/0.5–1 cm in diameter, conical to convex, cinnamon-brown becoming paler on drying, smooth, with striations at the margin. *Gills* adnexed, rusty brown, moderately crowded. *Stem* $\frac{3}{8}$–$1\frac{1}{4}$ × $\frac{1}{16}$ in/1–3.5 × 0.1–0.2 cm, cylindrical, cream-coloured becoming chestnut-brown below, bearing a conspicuous ring midway up the stem. *Flesh* thin, brown. *Spore deposit* rusty brown. *Habitat* on clay soil, often in pastures. *Edibility* **poisonous, stated to be "deadly".**

*CONOCYBE FILARIS*

**THREAD CONE CAP or DEADLY CONOCYBE**

GROUP Mushrooms and Toadstools

FAMILY Bolbitius (Bolbitiaceae)

SEASON EDIBILITY

SIMILAR SPECIES

One of the few cone caps with a ring on the stem; most of the related species are more robust.

## *PSATHYRELLA GRACILIS*

**A** tall, slender, brittle toadstool found growing in groups, and distinguished by the dark grey gills which have a pink edge.

*Cap* $\frac{3}{4}$–$1\frac{1}{4}$ in/2–3 cm in diameter, bell-shaped then expanded, brown when moist, drying to greyish brown, smooth, with a striated margin. *Gills* adnate, greyish to black, with a pink edge, narrow, moderately crowded. *Stem* $1\frac{1}{4}$–4 × $\frac{1}{8}$ in/3–10 × 0.2–0.3 cm, slender, tall, white, with a hairy base. *Flesh* thin, watery. *Spore deposit* purplish black. *Habitat* common in groups but not tufted, both in grasslands and woodlands. *Edibility* inedible.

*PSATHYRELLA GRACILIS*

**SLENDER PSATHYRELLA**

GROUP Mushrooms and Toadstools

FAMILY Ink-cap (Coprinaceae)

SEASON EDIBILITY

SIMILAR SPECIES

There are several hundred *Psathyrella* species, mostly recognized on the basis of microscopical characters, but this is one of the more common and more conspicuous species.

## *PSILOCYBE SEMILANCEATA*

**A** well-known species, owing to its reputation as a "magic mushroom".

*Cap* $\frac{3}{8}$–$\frac{5}{8}$ in/1–1.5 cm in diameter, narrowly conical with a central, pointed projection, not expanding, pale yellowish brown drying to almost white, smooth, sticky, with a darker, striated margin. *Gills* adnate, grey-brown to blackish brown, broad and crowded. *Stem* 2–3 × $\frac{1}{8}$ in/5–8 × 0.2–0.3 cm, slender, cylindrical, paler than the cap and often bruising bluish green towards the base. *Flesh* thin, firm. *Spore deposit* purplish black. *Habitat* very common, solitary or in very large numbers, in open grassland. *Edibility* toxic, causing psychotropic poisoning, and consequently has been used as an hallucinogen.

*PSILOCYBE SEMILANCEATA*

**LIBERTY CAP**

GROUP Mushrooms and Toadstools

FAMILY Stropharia (Strophariaceae)

SEASON EDIBILITY

SIMILAR SPECIES

There are numerous species of *Psilocybe*, and many are small and similar in appearance. The Bluing Psilocybe (*P. cyanescens*) lacks a point on the cap, while *P. fimetaria* grows on dung; both poisonous.

## STROPHARIA CORONILLA

**A** pale yellow mushroom, easily mistaken for the true mushrooms.

*Cap* ¾–2¼ in/2–6 cm in diameter, convex becoming flattened, pale yellow, smooth, sticky, with remains of the delicate veil hanging from the margin. *Gills* adnate, at first white becoming purplish brown, broad and crowded. *Stem* 1¼–2¼ × 3/16–⅜ in/3–6 × 0.5–1 cm, short, cylindrical, white, dry with a slightly woolly to dry surface, and bearing a narrow, white ring which has radial striations. *Flesh* thick, white, soft. *Spore deposit* dark purplish brown. *Habitat* in groups among short grass, including lawns. *Edibility* said to be poisonous and should be avoided.

*STROPHARIA CORONILLA*

**GARLAND STROPHARIA**

**GROUP** Mushrooms and Toadstools

**FAMILY** Stropharia (Strophariaceae)

SEASON EDIBILITY

SIMILAR SPECIES

The overall habit strongly resembles the true mushrooms (*Agaricus* species), but the gills are broadly attached to the stem and never free.

## INOCYBE PATOUILLARDII

**O**ne of the large species of *Inocybe*, distinguished by the red staining on handling; often found in the springtime.

*Cap* 1¼–2¾ in/3–7 cm in diameter, conical to bell-shaped, with a more or less pointed centre, white then yellowish, finely striated, staining reddish when handled. *Gills* adnexed, at first white soon darkening to olive-brown, crowded. *Stem* 1½–2¾ × 3/16–⅜ in/4–7 × 0.5–1 cm, cylindrical, robust, smooth, white, discolouring to red. *Flesh* white becoming reddish when broken open. *Spore deposit* cinnamon-brown. *Habitat* in groups in pastureland, also associated with beech trees. *Edibility* **poisonous and can be deadly, causing muscarine poisoning.**

*INOCYBE PATOUILLARDII*

**RED INOCYBE**

**GROUP** Mushrooms and Toadstools

**FAMILY** Cortinarius (Cortinariaceae)

SEASON EDIBILITY

SIMILAR SPECIES

*I. jurana* has a similar habit but a more pinkish brown cap and is found in both coniferous and beech woods; poisonous.

## CHALCIPORUS PIPERATUS

**A** small bolete recognized by the rusty tints on the pores and the bright yellow, tapering stem base.

*Cap* ¾–2¼ in/2–6 cm in diameter, convex then flattened, yellowish brown to cinnamon-brown, smooth, sticky when moist, with an incurved margin when young. *Tubes* adnate to subdecurrent, ochre-yellow, up to 1 cm deep; *pores* small, angular, yellowish to brick-red. *Stem* ¾–4 × ⅜–¾ in/2–10 × 1–2 cm, slender, cylindrical with a tapering base, yellowish brown with a bright yellow base. *Flesh* thick, pale yellow, with a hot, peppery taste. *Spore deposit* cinnamon-brown. *Habitat* prefers sandy soil, usually grows under birch and conifers. *Edibility* inedible, owing to peppery taste which can cause stomach upsets.

*CHALCIPORUS PIPERATUS*

**PEPPERY BOLETE**

**GROUP** Boletes

**FAMILY** Boletus (Boletaceae)

SEASON EDIBILITY

SIMILAR SPECIES

*Chalciporus* species are distinguished from the other boletes by the coppery red tints of the pore surface.

## CLAVARIA VERMICULARIS

The fungus forms a tuft of erect, pure white, slender clubs.

*Fruitbody* 2¼–4¾ in/6–12 cm high, comprising slender clubs, ⅛–3⁄16 in/0.3–0.5 cm wide, pure white and very brittle; the individual clubs taper at each end, and are sometimes tinted yellow at the tips. *Flesh* soft, white. *Spore deposit* white. *Habitat* among grass or on bare soil. *Edibility* said to be edible.

*CLAVARIA VERMICULARIS*

**WHITE SPINDLES or WHITE WORM CORAL FUNGUS**

**GROUP** Club Fungi

**FAMILY** Fairy Club (Clavariaceae)

**SEASON** **EDIBILITY**

**SIMILAR SPECIES**

*C. acuta* has a distinct base to the clubs which are also more delicate; inedible.

## CLAVULINOPSIS FUSIFORMIS

The fruitbody consists of tufts of bright yellow, slender clubs.

*Fruitbody* densely tufted, 2–5½ in/5–14 cm high, of bright yellow clubs which are fused at their base; individual clubs are 1⁄16–¼ in/0.2–0.6 cm wide, with a pointed tip. *Flesh* yellow, soft. *Spore deposit* white. *Habitat* among grass in field and woods. *Edibility* inedible, with a bitter taste.

*CLAVULINOPSIS FUSIFORMIS*

**GOLDEN SPINDLES**

**GROUP** Club Fungi

**FAMILY** Fairy Club (Clavariaceae)

**SEASON** **EDIBILITY**

**SIMILAR SPECIES**

*C. luteoalba* is also yellow but has white, obtusely rounded tips; *C. helvola* (see this page) has yellow but apically obtuse and non-tufted fruitbodies. Both species are inedible.

## CLAVULINOPSIS HELVOLA

This club-fungus differs from other yellow species in not being tufted.

*Fruitbody* consists of slender clubs, ¾–3 in/2–8 cm high and about ⅛ in/0.2–0.4 cm wide, cylindrical and unbranched, yellow to orange-yellow, with a more or less distinct base. *Flesh* whitish to yellow, soft and brittle. *Spore deposit* white. *Habitat* among grass and moss. *Edibility* inedible.

*CLAVULINOPSIS HELVOLA*

**GOLDEN MEADOW CLUB**

**GROUP** Club Fungi

**FAMILY** Fairy Club (Clavariaceae)

**SEASON** **EDIBILITY**

**SIMILAR SPECIES**

*C. fusiformis* (see this page) is yellow but has tufted pointed fruitbodies; inedible.

## *CLAVULINOPSIS CORNICULATA*

**A** branched, bright yellow club-fungus.

*Fruitbody* ¾–3 in/2–8 cm high, with numerous, slender clubs, branching two or three times with incurved tips, egg-yellow, bruising brownish, arising from a short, stalk-like base. *Flesh* rather tough. *Spore deposit* white. *Habitat* generally gregarious, on lawns and in woods. *Edibility* inedible.

*CLAVULINOPSIS CORNICULATA*

**COW HORN CLAVARIA**

**GROUP** Club Fungi

**FAMILY** Fairy Club (Clavariaceae)

**SEASON** **EDIBILITY**

**SIMILAR SPECIES**

Superficially similar to the inedible Jelly Antler Fungus (*Calocera viscosa*) but always growing on the ground, never on wood.

## *ALBATRELLUS OVINUS*

**O**ne of the few polypore fungi which grows on the ground, rather than on wood.

*Cap* 2¼–4 in/6–10 cm in diameter, strongly convex then flattened or depressed, often irregular, dry, white to yellowish, bruising brownish, smooth or with ill-defined small scales, and a wavy margin. *Tubes* decurrent, short, $\frac{1}{16}$ in/0.1–0.2 cm long, white; *pores* very small and difficult to see with the naked eye, white to lemon-yellow. *Stem* 1¼–2¼ × ⅜–¾ in/3–6 × 1–2 cm, central or excentric, curved, solid, hard, whitish. *Flesh* thick, white, with a fruity odour. *Spore deposit* white. *Habitat* among grass in open spruce forests, especially in mountainous areas. *Edibility* edible and good, with a firm consistency, and sometimes collected in very large quantities.

*ALBATRELLUS OVINUS*

**SHEEP POLYPORE**

**GROUP** Bracket fungi

**FAMILY** Hydnum (Hydnaceae)

**SEASON** **EDIBILITY**

**SIMILAR SPECIES**

*A. confluens* is pale orange, with a white pore surface; edible.

## *ASTRAEUS HYGROMETRICUS*

**A**n unusual earthstar which has star-like rays which open or close according to the weather conditions.

*Fruitbody* 1¼–2¾ in/3–7 cm in diameter, with a relatively large spore-sac and numerous curved rays. *Spore-sac* $\frac{1}{16}$–1¼ in/1–3 cm in diameter, almost ball-shaped, with a small torn pore at the apex, whitish to greyish brown, with the contents becoming powdery with age. *Rays* 6–12, up to 2 in/5 cm long, strongly curved up and over the spore-sac when dry but expanding and curving downwards in wet weather, greyish brown and cracked. *Spore deposit* brown at maturity. *Habitat* prefers sandy soil, often under conifers but uncommon. *Edibility* inedible.

*ASTRAEUS HYROMETRICUS*

**BAROMETER EARTHSTAR**

**GROUP** Puffballs and allies

**FAMILY** Earthstar Geastraceae

**SEASON** **EDIBILITY**

**SIMILAR SPECIES**

Most other earthstars (*Geastrum* species) lack rays which open and close.

### *BOVISTA NIGRESCENS*

One of the common, small, white puffballs, characterized by the absence of any stalk-like base.

*Fruitbody* 1¼–2¼ in/3–6 cm in diameter, ball-shaped, pure white but with the outer layer progressively flaking away to leave a shiny, black inner layer exposed, and the apex breaking open in an irregular fashion to release the powdery spore-mass. *Flesh* white and firm, then purplish black and powdery. *Spore deposit* purplish black. *Habitat* common, on open grassland, at first attached at the base but eventually breaking away and rolling free. *Edibility* edible when young and while the flesh is still white.

*BOVISTA NIGRESCENS*

**DUSTY PUFFBALL**

**GROUP** Puffballs and allies

**FAMILY** Puffball (Lycoperdaceae)

**SEASON** **EDIBILITY**

**SIMILAR SPECIES**

*B. plumbea* is smaller, with a greyish inner layer; edible.

### *LANGERMANNIA GIGANTEA*

Probably the easiest of all fungi to recognize, forming large ball-shaped structures, up to 63 in/160 cm in diameter.

*Fruitbody* 6–24 (–63) in/15–60 (–160) cm in diameter, ball-shaped or nearly so, rather flattened in large specimens, at first pure white, with a smooth, soft surface and a texture of suede, later discolouring yellowish to olive-brown with the outer layer gradually flaking away. *Flesh* white and firm but progressively becoming brown and powdery as the spores mature. *Spore deposit* brown. *Habitat* singly or in groups, sometimes forming large fairy rings, in field and on road verges. *Edibility* edible and delicious when young and the flesh is white and firm.

*LANGERMANNIA GIGANTEA*

**GIANT PUFFBALL**

**GROUP** Puffballs and allies

**FAMILY** Puffball (Lycoperdaceae)

**SEASON** **EDIBILITY**

**SIMILAR SPECIES**

No species grows to this size; *Calvatia excipuliformis* (see page 65) forms fruitbodies up to 4 in/10 cm in diameter, but has a basal stalk; edible.

### *CALVATIA UTRIFORMIS*

An occasional, large puffball, which may be recognized by the surface cracking into pyramid-shaped scales.

*Fruitbody* 2–6 in/5–15 cm in diameter, with a swollen but flattened upper region which tapers into a short, stout, stalk-like base, at first white and smooth soon cracking into pyramid-like scales which finally flake away. *Flesh* at first white and firm becoming powdery and olive-brown. *Spore deposit* olive-brown. *Habitat* among grass in meadows. *Edibility* edible when young and still white inside.

*CALVATIA UTRIFORMIS*

**MOSAIC PUFFBALL**

**GROUP** Puffballs and allies

**FAMILY** Puffball (Lycoperdaceae)

**SEASON** **EDIBILITY**

**SIMILAR SPECIES**

*C. cyathiformis* develops a purplish brown flesh and spore mass; edible.

## VASCELLUM PRATENSE

**A** small puffball frequently found on short grass, and distinguished by the pear-shaped fruitbody with short stalk, and a membrane separating the spore-mass from the base.

*Fruitbody* ¾–1½ in/2–4 cm in diameter, pear-shaped, tapering below into a stalk-like base, white to cream, covered with small granules and spines which are easily rubbed off, eventually developing a torn pore at the apex through which the spores escape; old worn fruitbodies resemble discarded cigar butts. *Flesh* white and firm becoming olive-brown and powdery. *Spore deposit* olive-brown. *Habitat* common in short grass. *Edibility* edible when young and white.

*VASCELLUM PRATENSE*

**LAWN PUFFBALL**

GROUP Puffballs and allies

FAMILY Puffball (Lycoperdaceae)

SEASON EDIBILITY

SIMILAR SPECIES

One of a number of similar puffballs, but differing from the *Lycoperdon* species by having the upper spore-sac separated from the stalk by a thin membrane.

## MORCHELLA ESCULENTA

**A** robust, club-shaped species recognized by the pale brownish, honeycomb-like head and white, scurfy stem.

*Cap* 2–4 in/5–10 cm high, 1½–2¼ in/4–6 cm wide, separated from the stalk by only a narrow groove, variable in shape, subglobose, oval or sometimes bluntly conical, yellowish brown, honeycomb-like, with irregular fertile pits separated by sterile ridges, hollow. *Stem* 1¼–2¼ × ¾–1½ in/3–6 × 2–4 cm, usually somewhat swollen towards the base, irregularly ribbed and furrowed, whitish, minutely scurfy, hollow. *Flesh* brittle, thin, whitish. *Spore deposit* ochraceous. *Habitat* open scrub or waste ground, often on chalky soil, sometimes in dune slacks. *Edibility* a much sought-after and excellent edible species, but should never be eaten raw.

*MORCHELLA ESCULENTA*

**COMMON MOREL or YELLOW MOREL**

GROUP Cup-fungi

FAMILY Morel (Morchellaceae)

SEASON EDIBILITY

SIMILAR SPECIES

*M. elata* (see this page) differs in having a narrower, conical cap with vertical, dark-coloured ribs; *M. hortensis* has a more elongated, rather conical cap and somewhat parallel but not dark-coloured ribs. Both edible.

## MORCHELLA ELATA

**A** woodland species recognized by the narrowly conical cap which has a honeycomb-like pattern of pale pits separated by blackish vertical ribs.

*Cap* ¾–2¼ in/2–6 cm high, ¾–1½ in/2–4 cm wide, separated from the stem by a narrow groove, hollow, narrowly conical, with regular, yellowish brown pits separated by dark grey or blackish vertical ribs and paler horizontal ribs. *Stem* 1½–3 × ¾–1½ in/4–8 × 2–4 cm, cylindrical or often slightly tapered downwards, hollow, whitish or cream, the surface coarsely mealy. *Flesh* thin, brittle, whitish. *Spore deposit* ochraceous. *Habitat* on soil in coniferous woodlands, sometimes on burnt ground. *Edibility* edible after cooking.

*MORCHELLA ELATA*

**BLACK MOREL**

GROUP Cup-fungi

FAMILY Morel (Morchellaceae)

SEASON EDIBILITY

SIMILAR SPECIES

*M. conica* has predominantly vertical, parallel ribs, but these are not darker coloured than the pits; edible.

## MORCHELLA SEMILIBERA

**A** distinctive, spring-fruiting species recognized by the pendant, conical, honeycombed cap, the lower half of which is free from the stem.

*Cap* ⅝–1½ in/1.5–4 cm high, ⅝–1¼ in/1.5–3 cm wide, conical, free from the stem for the lower half of its length, with yellowish brown fertile pits separated by irregular, mostly vertical, dark brown or blackish ribs. *Stem* 2¼–4 × ⅜–¾ in/6–10 × 1–2 cm, hollow, often tall, usually thickened towards the base, whitish to cream, surface scurfy. *Flesh* thin, brittle, whitish. *Habitat* damp soil in woods, in hedges or on waste ground. *Edibility* a good edible species, but should always be cooked.

*MORCHELLA SEMILIBRA*

**HALF FREE MOREL**

**GROUP** Cup-fungi

**FAMILY** Morel (Morchellaceae)

**SEASON** **EDIBILITY**

**SIMILAR SPECIES**

*M. elata* (see page 90) has dark ribs, but has the cap fully attached to the stem; edible.

## ALEURIA AURANTIA

**A** large, distinctive species recognized by the bright orange, usually clustered, cup-shaped fruitbodies.

*Fruitbody* ¾–4 in/2–10 cm across, irregularly cup-shaped, often expanded and sometimes split at the margin, lacking a stalk. *Inner surface* bright orange, smooth. *Outer surface* whitish or very pale orange, minutely downy, especially near the margin. *Flesh* thin, fragile, whitish, without a distinctive smell. *Habitat* usually on bare soil or in open woods, gregarious and often clustered. *Edibility* edible but worthless.

*ALEURIA AURANTIA*

**ORANGE PEEL FUNGUS**

**GROUP** Cup fungi

**FAMILY** Eyelash Cup Fungi (Pyronemataceae)

**SEASON** **EDIBILITY**

**SIMILAR SPECIES**

*Melastiza chateri* (see this page) is bright orange, but has smaller, less cup-shaped fruitbodies with short, brown hairs at the margin; inedible.

## MELASTIZA CHATERI

**A** bright orange, saucer-shaped species recognized by the bunches of minute brown hairs present near the margin of the underside of the fruitbody.

*Fruitbody* 3⁄16–¾ in/0.5–2 cm across, at first cup-shaped, later expanded, flattened, sometimes curved at the margin. *Inner surface* bright vermilion-orange, smooth. *Outer surface* paler, bearing tiny bunches of brown hairs, especially at the margin. *Flesh* pale orange. *Habitat* on bare, damp, usually sandy soil. *Edibility* inedible.

*MELASTIZA CHATERI*

**CHATER'S VERMILION CUP**

**GROUP** Cup-fungi

**FAMILY** Eyelash Cup Fungi (Pyronemataceae)

**SEASON** **EDIBILITY**

**SIMILAR SPECIES**

Other species of *Melastiza* may be similar and can be distinguished only by microscopic examination. *Aleuria aurantia* (see this page) is larger and lacks brown hairs; edible.

## *TRICHOGLOSSUM HIRSUTUM*

**A** blackish, erect, club-shaped species distinguished by the velvety surface of the fruitbody.

*Fruitbody* 1¼–3 in/3–8 cm high, black or blackish brown, velvety throughout due to minute, stiff, blackish hairs, variable in shape, narrowly club-shaped or sometimes capitate, upper fertile part ⅛–¼ in/0.3–0.6 cm wide, laterally compressed, irregularly furrowed, sharply delimited from the stem. *Stem* up to 2¼ in/6 cm high, ⅛ in/0.2–0.4 cm wide, slightly compressed and sometimes furrowed. *Flesh* dark brown. *Habitat* usually in wet acid grassland, often among *Sphagnum* moss. *Edibility* edible.

*TRICHOGLOSSUM HIRSUTUM*

**HAIRY EARTH TONGUE**

**GROUP** Cup-fungi

**FAMILY** Earth Tongue (Geoglossaceae)

**SEASON** **EDIBILITY**

**SIMILAR SPECIES**

Other species of *Trichoglossum* can be distinguished only by microscopic examination. Species of *Geoglossum* are similar in shape and colour, but lack the velvety surface.

## *MICROGLOSSUM VIRIDE*

**An** erect, club-shaped species with greenish fruitbodies and characterized by the scurfy-scaly stem.

*Fruitbody* ¾–2¼ in/2–6 cm high, narrowly clavate, green or olive-green throughout, upper fertile part smooth, furrowed, somewhat compressed, ⅛–¼ in/0.3–0.7 cm wide, sharply delimited from the stem. *Stem* 1⁄16–3⁄16 in/0.2–0.5 cm thick, often slightly compressed, cylindrical, often curved, the surface scurfy or scaly. *Flesh* pale greenish. *Habitat* on soil in woodland, often among mosses; usually gregarious, sometimes clustered. *Edibility* inedible.

*MICROGLOSSUM VIRIDE*

**GREEN EARTH TONGUE**

**GROUP** Cup-fungi

**FAMILY** Earth Tongue (Geoglossaceae)

**SEASON** **EDIBILITY**

**SIMILAR SPECIES**

*M. olivaceum* is similar in form, but differs in the brownish colour of the fertile head and in having a smooth stalk; inedible.

## *CORDYCEPS MILITARIS*

**A** distinctive, club-shaped species recognized by the bright orange fruitbody with roughened head and smooth stem which arises from a buried, mummified caterpillar.

*Fruitbody* ¾–2¼ in/2–6 cm high, cylindrical-clavate, upper fertile part 3⁄16–¼ in/0.4–0.6 cm wide, bright orange or orange-red, cylindrical or irregular in shape, with a finely roughened surface. *Stem* ⅛–3⁄16 in/0.3–0.4 cm wide, cylindrical, flexuous, smooth, orange, slightly paler than the head. *Habitat* in grassland, arising singly or in clusters from buried, mummified larvae or pupae of butterflies or moths. *Edibility* inedible.

*CORDYCEPS MILITARIS*

**TROOPING CORDYCEPS or SCARLET CATERPILLAR FUNGUS**

**GROUP** Flask-fungi

**FAMILY** Vegetable Caterpillar (Clavicipitaceae)

**SEASON** **EDIBILITY**

**SIMILAR SPECIES**

Unlikely to be confused with other species.

CHAPTER FOUR

# On Trees, Stumps or Woody Debris

## PLEUROTUS OSTREATUS

**A** well-known edible species, growing in tiers on trunks and stumps of frondose trees, with a greyish brown cap and lateral attachment.

*Cap* 2–5½ in/5–14 cm in diameter, shell-shaped, semi-circular, gradually expanding and becoming flattened with a wavy margin, whitish to greyish brown, blue-grey or even deep violet, smooth and dry. *Gills* decurrent, white, broad, and moderately crowded. *Stem* none, or very short and laterally attached, thick and solid. *Flesh* thick, white, with a pleasant "mushroom" odour. *Spore deposit* very pale lilac. *Habitat* on stumps and trunks of frondose trees, especially beech. *Edibility* excellent; often grown and sold commercially.

*PLEUROTUS OSTREATUS*

**OYSTER MUSHROOM**

**GROUP** Bracket-fungi

**FAMILY** Lentinus and Oyster Cap (Lentinaceae)

**SEASON** **EDIBILITY**

**SIMILAR SPECIES**

There are many varieties, mostly based upon cap colour. P. pulmonarius is thinner and has a whitish cap, while the Branched Oyster Mushroom (*P. cornucopiae*) has gills which branch and form a network over the top of the stem; both edible.

## PLEUROTUS DRYINUS

**A** large oyster mushroom, usually not tufted, distinguished by the ring on the stem and the yellowing gills.

*Cap* 2¾–5½ in/7–14 cm in diameter, strongly convex to shell-shaped, whitish, with a felty surface, dry and sometimes cracking, with fragments of the veil hanging from the margin. *Gills* decurrent, white drying yellowish, broad, moderately crowded. *Stem* ¾–2 × ⅜–1¼ in/2–5 × 1–3 cm, off-centre, short, whitish, solid, with a thin, membranous, greyish ring attached to the upper region which rapidly disintegrates. *Flesh* thick, white, with an aromatic odour. *Spore deposit* white. *Habitat* on deciduous trees. *Edibility* edible when young but developing a firm to hard texture.

*PLEUROTUS DRYINUS*

**VEILED OYSTER**

**GROUP** Bracket-fungi

**FAMILY** Lentinus and Oyster Cap (Lentinaceae)

**SEASON** **EDIBILITY**

**SIMILAR SPECIES**

Other oyster mushrooms form clusters, and lack a ring on the stem.

## PLEUROCYBELLA PORRIGENS

**O**ften forms large numbers of white clusters, with fan-shaped fruitbodies, on coniferous stumps and debris.

*Cap* ¾–4 in/2–10 cm in diameter, fan-shaped, often erect, pure white, dry and smooth, with an incurved margin. *Gills* decurrent, creamy white, narrow, and very crowded. *Stem* none or very reduced. *Flesh* thin, white, brittle. *Spore deposit* white. *Habitat* on rotting coniferous wood, prefers colder regions. *Edibility* edible.

*PLEUROCYBELLA PORRIGENS*

**ANGEL'S WINGS**

**GROUP** Mushrooms and Toadstools

**FAMILY** Tricholoma (Tricholomataceae)

**SEASON** **EDIBILITY**

**SIMILAR SPECIES**

This species is thinner-fleshed than most of the true oyster mushrooms.

## *LENTINELLUS COCHLEATUS*

Forms crowded tufts on the wood of frondose trees, and dries without rotting.

*Cap* ¾–2¾ in/2–7 cm in diameter, fan- to funnel-shaped, erect, reddish brown, dry and smooth, with a thin, lobed margin. *Gills* decurrent, whitish developing a pinkish tint, narrow, very crowded, with a saw-like edge. *Stem* ¾–4 × 3/16–⅜ in/2–10 × 0.5–1 cm, central or off-centre, short and usually fusing at the base with other stems to form a cluster, rusty brown. *Flesh* thin, whitish, easily splitting, with an odour of aniseed. *Spore deposit* white. *Habitat* on the old branches, stumps and buried wood of deciduous trees. *Edibility* said to be edible when young.

*LENTINELLUS COCHLEATUS*

**COCKLESHELL LENTINUS**

**GROUP** Mushrooms and Toadstools

**FAMILY** Lentinus and Oyster Cap (Lentinaceae)

**SEASON** **EDIBILITY**

**SIMILAR SPECIES**

*L. omphalodes* is a whitish species, with a slender, central stem, which does not form clusters; inedible.

## *SCHIZOPHYLLUM COMMUNE*

A common species of warmer regions, forming tiers of small, grey, hairy brackets on stumps and trunks.

*Cap* ⅜–1½ in/1–4 cm in diameter, shell-shaped and laterally attached, pale grey or pure white in very dry conditions, densely hairy-scaly. *Gills* radiating from a lateral attachment point, appearing to split lengthwise along their edges in dry weather and the sides curling upwards, narrow, grey. *Stem* none. *Flesh* thin, grey, leathery. *Spore deposit* white. *Habitat* on dead branches of deciduous trees, especially beech, also on domestic wood. *Edibility* inedible, too tough.

*SCHIZOPHYLLUM COMMUNE*

**SPLIT GILL**

**GROUP** Mushrooms and Toadstools

**FAMILY** Split-gill (Schizophyllaceae)

**SEASON** **EDIBILITY**

**SIMILAR SPECIES**

The splitting gills and the tough, leathery texture separate this species from all other oyster mushrooms.

## *ARMILLARIA MELLEA*

A common, fleshy mushroom, growing in large tufts on stumps and among tree roots; recognized by the whitish gills and the ring on the stem. Honey fungus is a serious disease of garden trees and shrubs.

*Cap* 1¼–4¾ in/3–12 cm in diameter, convex to flattened with a wavy margin, yellow-brown (honey colour), with tiny, scattered dark brown scales, sticky when young and fresh. *Gills* shortly decurrent, whitish but developing reddish stains, narrow and moderately crowded. *Stem* 2–6 × ⅜–¾ in/5–15 × 1–2 cm, cylindrical, soon hollow, whitish becoming rusty brown, fibrous, with a thick, cottony ring attached to the upper region; the stem is attached at the base to thick, black, coarse threads which branch and spread over the host plant and through the soil. *Flesh* white, firm, with a bitter taste. *Spore deposit* white. *Habitat* clustered on roots and stumps of deciduous trees. *Edibility* caps edible when young and fresh, but allergic reactions shown by some.

*ARMILLARIA MELLEA*

**HONEY FUNGUS**

**GROUP** Mushrooms and Toadstools

**FAMILY** Tricholoma (Tricholomataceae)

**SEASON** **EDIBILITY**

**SIMILAR SPECIES**

*A. tabescens* lacks a ring on the stem, while *A. ostoyae* has a pinkish cap and attacks conifer trees. Edibility in both species is as for *A. mellea* above.

## TRICHOLOMOPSIS DECORA

**A** distinctive species on coniferous stumps, especially in mountainous regions, with a yellowish cap with dark brown scales.

*Cap* 1¼–3 in/3–8 cm in diameter, convex becoming depressed in the centre, pale yellowish orange, with tiny, blackish brown, fibrous scales over the surface, especially towards the centre. *Gills* adnexed, yellow, narrow and crowded. *Stem* 1¼–2¼ × 3/16–3/8 in/3–6 × 0.5–1 cm, cylindrical, yellow, with small, darker scales. *Flesh* whitish, thick. *Spore deposit* white. *Habitat* on stumps and dead wood of coniferous trees. *Edibility* inedible, and unpleasant.

*TRICHOLOMOPSIS DECORA*

**DECORATED MOP**

GROUP Mushrooms and Toadstools

FAMILY Tricholoma (Tricholomataceae)

SEASON EDIBILITY

SIMILAR SPECIES

*T. rutilans* (see this page) has purplish tints, while the Yellow Oyster Mop (*T. sulphuroides*) is paler yellow with dark radiating streaking; both inedible.

## TRICHOLOMOPSIS RUTILANS

**A** common, brightly coloured toadstool found on conifer stumps, with yellow and purple colours.

*Cap* 2–6 in/5–15 cm in diameter, convex finally flattened, often with a raised centre, pale yellowish but densely covered with tiny, dark reddish scales, and with an inrolled margin. *Gills* adnexed, yellow, narrow to broad, crowded. *Stem* 2–4 × 3/8–3/4 in/5–10 × 1–2 cm, cylindrical, often curved, yellowish with reddish scales, becoming hollow. *Flesh* thick, white, firm. *Spore deposit* white. *Habitat* often in small groups, especially on dead pine wood. *Edibility* inedible, with bitter taste.

*TRICHOLOMOPSIS RUTILANS*

**PURPLE AND YELLOW AGARIC or VARIEGATED MOP**

GROUP Mushrooms and Toadstools

FAMILY Tricholoma (Tricholomataceae)

SEASON EDIBILITY

SIMILAR SPECIES

*T. decora* (see this page) is yellow, with small brown scales, but lacks any purplish tints; inedible.

## MARASMIELLUS RAMEALIS

**A** tiny, pale, wrinkled toadstool, forming dense clusters on bramble stems.

*Cap* 3/16–3/8 in/0.5–1 cm in diameter, thin, convex soon flattened, often appearing wrinkled, white with a pinkish centre, dry. *Gills* few, adnate, white, narrow and well spaced. *Stem* 3/16–3/8 × 1/16 in/0.5–1 × 0.1–0.2 cm, often curved, white becoming reddish brown from the base, smooth. *Flesh* thin, membranous, white. *Spore deposit* white. *Habitat* clusters on old, fallen twigs, especially bramble stem. *Edibility* inedible.

*MARASMIELLUS RAMEALIS*

**TWIG MARASMIUS**

GROUP Mushrooms and Toadstools

FAMILY Tricholoma (Tricholomataceae)

SEASON EDIBILITY

SIMILAR SPECIES

The White Marasmius (*M. candidus*) has a grooved cap and pale pinkish gills; inedible.

## *COLLYBIA FUSIPES*

**A** dark brown species that has a swollen stem and forms tufts at the base of beech and oak trees.

*Cap* $1\frac{1}{4}$–4 in/3–10 cm in diameter, bell-shaped or with a raised centre, moist, tawny brown becoming paler from the centre outwards, smooth and shiny. *Gills* adnexed, pale grey, often spotted with rusty brown marks, broad, widely spaced. *Stem* 3–$4\frac{3}{4}$ × $\frac{3}{8}$–$\frac{3}{4}$ in/8–12 × 1–2 cm, flattened and ridged, smooth, tapering at the base, at first whitish but soon discolouring dark reddish brown from the base up; the stems are fused into a root-like structure. *Flesh* thin but tough, white. *Spore deposit* white. *Habitat* clustered among the roots of beech and oak trees. *Edibility* inedible.

| *COLLYBIA FUSIPES* |
|---|
| **SPINDLE SHANK** |
| GROUP Mushrooms and Toadstools |
| FAMILY Tricholoma (Tricholomataceae) |
| SEASON · EDIBILITY |
| SIMILAR SPECIES |
| The Greasy Tough Shank (*C. butyracea*, page 28) has a swollen stem, but it does not grow in clusters; inedible. |

## *MICROMPHALE FOETIDUM*

**A** small, thin-fleshed, dark brown toadstool, growing in clusters on fallen twigs, and distinguished by the velvety stem and a very unpleasant odour.

*Cap* $\frac{3}{8}$–$1\frac{1}{2}$ in/1–4 cm in diameter, thin, convex with a depressed centre, soon becoming flattened with a wavy margin, dark reddish brown, with radiating grooves. *Gills* adnate, yellow but developing brown tints, well spaced. *Stem* $\frac{3}{4}$–$1\frac{1}{2}$ × $\frac{1}{16}$ in/2–4 × 0.1–0.2 cm, wiry but hollow, tapering towards the base, dark brown, with velvety hairs. *Flesh* membranous, brown, with a very unpleasant odour recalling rotting fish. *Spore deposit* white. *Habitat* on fallen, woody debris, growing especially on beech and hazel. *Edibility* inedible.

| *MICROMPHALE FOETIDUM* |
|---|
| **FETID MARASMIUS** |
| GROUP Mushrooms and Toadstools |
| FAMILY Tricholoma (Tricholomataceae) |
| SEASON · EDIBILITY |
| SIMILAR SPECIES |
| *M. brassicolens* grows on beech leaves and fallen twigs, with a cabbage-water smell; inedible. |

## *FLAMMULINA VELUTIPES*

**A** tufted mushroom, growing on trees during the winter months, recognized by the velvety stem and slimy cap.

*Cap* $1\frac{1}{4}$–$2\frac{1}{4}$ in/3–6 cm in diameter, convex but soon flattened with a raised centre, reddish brown, sticky to slimy, smooth. *Gills* adnexed, white, broad, moderately spaced. *Stem* $1\frac{1}{4}$–$2\frac{3}{4}$ × $\frac{1}{8}$–$\frac{3}{16}$ in/3–7 × 0.3–0.5 cm, tough, tapering at the base, yellowish above becoming blackish brown below, finely velvety. *Flesh* soft, white. *Spore deposit* white. *Habitat* tufted on deciduous trees, especially elm. *Edibility* edible and collected throughout winter.

| *FLAMMULINA VELUTIPES* |
|---|
| **VELVET SHANK** |
| GROUP Mushrooms and Toadstools |
| FAMILY Tricholoma (Tricholomataceae) |
| SEASON · EDIBILITY |
| SIMILAR SPECIES |
| This species is grown commercially and sold under the name of "Eno-take"; however, the cultivated variety looks very different tending to be white, with a tiny cap, and very densely tufted. |

## *MEGACOLLYBIA PLATYPHYLLA*

**A** large mushroom characterized by the grey, streaky cap, and the white mycelial cords attached to the stem base.

*Cap* 2–4 in/5–10 cm in diameter, convex soon becoming flattened, greyish brown, dry, with conspicuous radial streaks. *Gills* adnexed, white, often very broad and easily splitting, well spaced. *Stem* 3–4¾ × ⅜–¾ in/8–12 × 1–2 cm, cylindrical and robust, white, fibrous, arising from conspicuous, white, branching mycelial cords. *Flesh* thin, white. *Spore deposit* white. *Habitat* on dead fallen wood and stumps of deciduous trees. *Edibility* not recommended, with a bitter taste and can cause stomach upsets.

*MEGACOLLYBIA PLATYPHYLLA*

**BROAD-GILLED AGARIC or PLATTERFUL MUSHROOM**

**GROUP** Mushrooms and Toadstools

**FAMILY** Tricholoma (Tricholomataceae)

**SEASON** **EDIBILITY**

**SIMILAR SPECIES**

This species differs from *Tricholoma* species by always growing on wood.

## *OUDEMANSIELLA MUCIDA*

**S**mall tufts of white toadstools, with very slimy caps, and a ring on the stem, often found on the upper branches of beech trees. Also known as "Poached Egg Fungus" because of the shiny, slimy caps.

*Cap* 1¼–2¾ in/3–7 cm in diameter, strongly convex, white or very pale grey, very slimy and almost transparent, becoming wrinkled on drying. *Gills* adnate, white, very broad and well spaced. *Stem* 2–3 × $^{3}/_{16}$–1¼ in/5–8 × 0.5–0.7 cm, curved, cylindrical, tough, smooth, with a white, membranous ring attached to the upper region. *Flesh* white and very soft. *Spore deposit* white. *Habitat* small tufts on branches, trunks and stumps of beech trees. *Edibility* edible but worthless.

*OUDEMANSIELLA MUCIDA*

**SLIMY BEECH CAPS or POACHED EGG FUNGUS**

**GROUP** Mushrooms and Toadstools

**FAMILY** Tricholoma (Tricholomataceae)

**SEASON** **EDIBILITY**

**SIMILAR SPECIES**

Unlikely to be confused with other species.

## *OUDEMANSIELLA RADICATA*

**A** large, common and conspicuous mushroom found in the autumn, with a shiny, brown cap and a deeply rooting stem.

*Cap* 2–4 in/5–10 cm in diameter, convex soon flattened and forming an upturned, wavy margin, yellowish brown to dark brown, slimy when moist, with a wrinkled centre. *Gills* adnate, white, thick and broad, moderately spaced. *Stem* 2–8 × $^{3}/_{16}$–⅜ in/5–20 × 0.5–1 cm, rather brittle, white at the apex, brownish below and expanding slightly downwards to form a long, underground root-like extension. *Flesh* thick, white. *Spore deposit* white. *Habitat* apparently growing on the ground but attached to underground tree roots, especially of beech. *Edibility* edible.

*OUDEMANSIELLA RADICATA*

**ROOTING SHANK**

**GROUP** Mushrooms and Toadstools

**FAMILY** Tricholoma (Tricholomataceae)

**SEASON** **EDIBILITY**

**SIMILAR SPECIES**

*O. pudens* is less common and has a dry, velvety cap and stem; inedible.

## *MARASMIUS ANDROSACEUS*

**A** tiny toadstool, growing among fallen conifer needles, along with additional black threads which do not develop caps.

*Cap* 3/16–3/8 in/0.5–1 cm in diameter, strongly convex with a depressed centre, pale reddish brown, wrinkled, dry. *Gills* adnate to almost decurrent, white, narrow, crowded. *Stem* 1¼–2¼ × 1/16 in/3–6 × 0.1–0.2 cm, black, wiry, smooth; the mycelium also forms thin, wiry cords resembling horse-hairs which entangle the dead pine needles. *Flesh* membranous. *Spore deposit* white. *Habitat* grows among conifer needles. *Edibility* edible but worthless.

*MARASMIUS ANDROSACEUS*

### HORSEHAIR TOADSTOOL

GROUP Mushrooms and Toadstools

FAMILY Tricholoma (Tricholomataceae)

SEASON EDIBILITY

SIMILAR SPECIES

*Micromphale perforans* is similar in appearance, but each fruitbody grows on a separate needle, the stem is velvety, and there is an unpleasant smell on bruising; inedible.

## *MARASMIUS ROTULA*

**A** tiny, membranous toadstool, growing in large numbers on fallen twigs, with widely spaced gills joined by a collar.

*Cap* 3/16–3/8 in/0.5–1 cm in diameter, strongly convex with a sunken centre, white to pale yellowish brown, smooth and dry, with broad, radial grooves. *Gills* broadly attached to a circular collar around the stem apex, white, widely spaced. *Stem* ¾–1½ × 1/16 in/2–4 × 0.1–0.2 cm, wiry, at first yellow soon blackish brown from the base upwards, smooth. *Flesh* membranous, white. *Spore deposit* white. *Habitat* growing in clusters on fallen twigs of deciduous trees. *Edibility* edible but worthless.

*MARASMIUS ROTULA*

### LITTLE WHEEL TOADSTOOL or PINWHEEL MARASMIUS

GROUP Mushrooms and Toadstools

FAMILY Tricholoma (Tricholomataceae)

SEASON EDIBILITY

SIMILAR SPECIES

The Leaf Marasmius (*M. epiphyllus*) lacks a collar at the stem apex, while *M. capillaripes* grows on oak leaves; both inedible.

## *STROBILURUS TENACELLUS*

**One** of the small toadstools which grow from buried conifer cones, this species appears in spring on pine cones.

*Cap* 3/8–3/4 in/1–2 cm in diameter, conical then expanding, at first whitish but soon yellowish red or brown, smooth and shiny. *Gills* adnexed, white, narrow, crowded. *Stem* ¾–2¾ × 1/16 in/2–7 × 0.1–0.2 cm, slender, cylindrical, half-buried, buffy brown but paler towards the top. *Flesh* thin, white. *Spore deposit* white. *Habitat* growing from buried pine cones. *Edibility* inedible.

*STROBILURUS TENACELLUS*

### SPRING PINE CONE TOADSTOOL

GROUP Mushrooms and Toadstools

FAMILY Tricholoma (Tricholomataceae)

SEASON EDIBILITY

SIMILAR SPECIES

*S. esculentus* grows from buried spruce cones, never pine, and occurs in the autumn months. *Baeospora myosura* also grows on pine cones but has a darker brown cap and yellowish gills. Both species are inedible.

## PANELLUS STIPTICUS

**A** small, brown, tough oyster mushroom, growing in tiers on dead wood, with very crowded narrow gills, which emit a greenish light in the dark.

*Cap* ⅜–1¼ in/1–3 cm in diameter, kidney-shaped, pale buffy brown, dry, scurfy to hairy. *Gills* adnate, pinkish brown, very narrow, crowded. *Stem* 3⁄16–⅝ × 3⁄16–⅜ in/0.5–1.5 × 0.5–1 cm, very short, strongly off-centre to lateral, broadest at the point of attachment to the cap, flattened, very pale brown, hairy. *Flesh* thin, tough, brown, with a bitter taste. *Spore deposit* white. *Habitat* clustered, on logs and stumps. *Edibility* inedible, bitter and possibly toxic.

*PANELLUS STIPTICUS*

**STYPIC FUNGUS or LUMINESCENT PANELLUS**

**GROUP** Mushrooms and Toadstools

**FAMILY** Tricholoma (Tricholomataceae)

**SEASON** **EDIBILITY**

**SIMILAR SPECIES**

This species is easily confused with *Crepidotus* species, which have brown gills and spores, and a spore print is required to be certain of identification.

## MYCENA GALERICULATA

**A** tufted elf cap, commonly found growing on rotting wood, with a bell-shaped brown cap and whitish gills which are interveined. This is one of the largest and more robust of the elf caps.

*Cap* ¾–2 in/2–5 cm in diameter, bell-shaped with a raised centre and then developing a slightly expanding edge, greyish brown, at times wrinkled. *Gills* adnexed, white or with a greyish pink tint, broad, well spaced but conspicuously cross-veined. *Stem* 1¼–2¾ × 1⁄16–3⁄16 in/3–7 × 0.2–0.5 cm, pale grey, more brownish towards the base, smooth and hollow. *Flesh* thin, white, with an odour of damp flour. *Spore deposit* white. *Habitat* tufted on rotting wood of deciduous trees. *Edibility* worthless.

*MYCENA GALERICULATA*

**BONNET MYCENA or COMMON MYCENA**

**GROUP** Mushrooms and Toadstools

**FAMILY** Tricholoma (Tricholomataceae)

**SEASON** **EDIBILITY**

**SIMILAR SPECIES**

*M. inclinata* (see this page) has reddish brown stem bases and a distinct smell.

## MYCENA INCLINATA

**A** densely tufted elf cap, growing on oak-stumps, with a characteristic reddish brown stem base.

*Cap* ⅜–1½ in/1–4 cm in diameter, bell-shaped, dark yellowish brown, smooth, radially striated, with a toothed margin. *Gills* adnate, whitish, narrow, crowded. *Stem* 2–3 × 1⁄16–⅛ in/5–8 × 0.2–0.3 cm, cylindrical, at first uniformly greyish and then darkening to yellow and finally dark reddish brown towards the base. *Flesh* thin, brittle, with an odour of rancid oil. *Spore deposit* white. *Habitat* tufted on oak-stumps. *Edibility* inedible.

*MYCENA INCLINATA*

**GREGARIUS ELF CAP**

**GROUP** Mushrooms and Toadstools

**FAMILY** Tricholoma (Tricholomataceae)

**SEASON** **EDIBILITY**

**SIMILAR SPECIES**

This species differs from the inedible Bonnet Mycena (*M. galericulata*, see this page) in the generally less robust habit and the yellowish to red-brown stem base.

## *MYCENA HAEMATOPUS*

**A** dark, clustered elf cap, growing on rotting wood, which releases a dark red juice when the stem is broken.

*Cap* ¾–1¼ in/2–3 cm in diameter, bell-shaped to convex, dark reddish brown, with radiating streaks, dry. *Gills* adnexed, whitish, staining reddish brown, moderately spaced. *Stem* 1¼–3 × 1/16–⅛ in/3–8 × 0.2–0.3 cm, cylindrical or flattened, reddish brown, with a hairy base. *Flesh* thin, reddish, which in the stem releases a dark blood-red liquid when broken. *Spore deposit* white. *Habitat* common growing on rotting wood. *Edibility* worthless.

*MYCENA HAEMATOPUS*

**BLEEDING MYCENA**

**GROUP** Mushrooms and Toadstools

**FAMILY** Tricholoma (Tricholomataceae)

**SEASON** **EDIBILITY**

**SIMILAR SPECIES**

The Small Bleeding Mycena (*M. sanguinolenta*) also releases a red juice when broken but grows on the ground, among moss, and is much more slender; inedible.

## *PLUTEUS CERVINUS*

**O**ne of the most common toadstools, recognized by the crowded, pink gills which are free from the stem apex.

*Cap* 1¼–4¾ in/3–12 cm in diameter, convex soon becoming flattened, greyish brown or darker, smooth but with radiating, darker striations, dry. *Gills* free, white but becoming salmon-pink as the spores develop, thin, very crowded. *Stem* 2–4 × 3/16–⅜ in/5–10 × 0.5–1 cm, cylindrical, solid, white, with fine, blackish brown fibres. *Flesh* thin, white. *Spore deposit* salmon-pink. *Habitat* common, on old stumps, fallen wood and sawdust. *Edibility* said to be edible when very young but soon decaying, not recommended.

*PLUTEUS CERVINUS*

**FAWN PLUTEUS**

**GROUP** Mushrooms and Toadstools

**FAMILY** Pluteus (Pluteaceae)

**SEASON** **EDIBILITY**

**SIMILAR SPECIES**

In North America, *P. magnus* is stouter; *P. abromarginatus* has a dark gill edge.

## *PLUTEUS UMBROSUS*

**T**he brown, velvety appearance of the cap and stem, and the pink gills readily distinguish this species.

*Cap* 2–4 in/5–10 cm in diameter, convex with a raised centre, uniformly brown but covered with minute, tiny, black scales, giving a velvety appearance. *Gills* free, white soon becoming pink, broad and crowded, with brown spots on the edge. *Stem* 1½–2¼ × ⅜–⅝ in/4–6 × 1–1.5 cm, cylindrical, solid, with a surface similar to the cap. *Flesh* pale, soft and soon rotting, with a faint garlic odour. *Spore deposit* salmon-pink. *Habitat* on dead wood of deciduous trees. *Edibility* edible, but tastes quite unpleasant.

*PLUTEUS UMBROSUS*

**BROWN PLUTEUS**

**GROUP** Mushrooms and Toadstools

**FAMILY** Pluteus (Pluteaceae)

**SEASON** **EDIBILITY**

**SIMILAR SPECIES**

*P. podospileus* has dark brown scales on the stem, but the gill edge is not spotted; inedible.

## PLUTEUS AURANTIORUGOSUS

One of the brightly coloured *Pluteus* species, distinguished by the scarlet-red cap.

*Cap* ¾–2¼ in/2–6 cm in diameter, conical then expanded with a raised centre, scarlet with an orange margin, smooth or wrinkled in the centre. *Gills* free, whitish becoming salmon-pink, crowded. *Stem* ¾–2¼ × 3⁄16–⅜ in/2–6 × 0.5–1 cm, cylindrical, whitish above becoming yellowish orange below, fibrous. *Flesh* thin, whitish, brittle. *Spore deposit* pink. *Habitat* on decayed wood, especially elm, ash or willow. *Edibility* edible.

*PLUTEUS AURANTIORUGOSUS*

**ORANGE PLUTEUS or GOLDEN GRANULAR PLUTEUS**

**GROUP** Mushrooms and Toadstools

**FAMILY** Pluteus (Pluteaceae)

**SEASON** **EDIBILITY**

**SIMILAR SPECIES**

The Yellow Pluteus (*P. leoninus*) is golden yellow, while *P. lutescens* has a brown cap and gills at first chrome-yellow; both inedible.

## VOLVARIELLA BOMBYCINA

An uncommon but beautiful species with silky fibres on the cap, pink gills, and a large sac-like volva.

*Cap* 2–8 in/5–20 cm in diameter, bell-shaped, whitish to pale yellowish, densely covered by fine, silky fibres. *Gills* free, pink, broad, and very crowded. *Stem* 2¾–4 × 3⁄16–⅜ in/7–10 × 0.5–1 cm, cylindrical, white, smooth and shiny, emerging from a basal sac-like, whitish volva with a brown-spotted surface. *Flesh* thin, soft, white, with a woody odour. *Spore deposit* salmon-pink. *Habitat* on dead elm-trunks, sometimes found high above the ground. *Edibility* edible but of poor quality.

*VOLVARIELLA BOMBYCINA*

**SILKY VOLVARIA or TREE VOLVARIA**

**GROUP** Mushrooms and Toadstools

**FAMILY** Pluteus (Pluteaceae)

**SEASON** **EDIBILITY**

**SIMILAR SPECIES**

The edible Rose-gilled Grisette (*V. speciosa*) has a sticky whitish cap and grows on enriched soil. The pink gills distinguish the *Volvariella* species from those of *Amanita*.

## RHODOTUS PALMATUS

Growing on rotting wood, usually elm, this species is distinguished by the tough, wrinkled cap surface and the pinkish gills.

*Cap* 1½–3 in/4–8 cm in diameter, strongly convex, pale pinkish orange, with a gelatinous, often strongly wrinkled surface, and an inrolled margin. *Gills* sinuate, yellowish becoming progressively pink, crowded. *Stem* ¾–2 × 3⁄16–⅜ in/2–5 × 0.5–1 cm, often off-centre and curved, fibrous, pale pinkish. *Flesh* thick, pale, with a fruity odour. *Spore deposit* pink. *Habitat* clusters on old stumps, especially elm and maple. *Edibility* inedible.

*RHODOTUS PALMATUS*

**NETTED RHODOTUS**

**GROUP** Mushrooms and Toadstools

**FAMILY** Pluteus (Pluteaceae)

**SEASON** **EDIBILITY**

**SIMILAR SPECIES**

The wrinkled cap is distinctive, and there are no closely related species.

## *PSATHYRELLA HYDROPHILA*

**G**rows in tufts on rotten wood, and is distinguished by the hygrophanous, brown, convex cap (which has a fringed margin), brown gills and contrasting white stem.

*Cap* 3/4–2 in/2–5 cm across, expanding, smooth, dark brown to chestnut when fresh, drying markedly paler from the centre; pale ochre when dry. Margin with white, tooth-like remnants of the veil. *Gills* brown, adnate, crowded. *Stem* 2–2 3/4 × 1/8–3/16 in/5–7 × 0.3–0.5 cm, white, smooth or with veil fibrils in the upper part. *Flesh* whitish, thin, fragile, hollow in the stem, with slightly bitter taste and no distinctive smell. *Spore deposit* dark brown. *Habitat* in dense clusters on stumps and logs of deciduous trees; common. *Edibility* inedible.

*PSATHYRELLA HYDROPHILA*

**WATERY HYPHOLOMA or CLUSTERED PSATHYRELLA**

**GROUP** Mushrooms and Toadstools

**FAMILY** Ink-cap (Coprinaceae)

**SEASON** **EDIBILITY**

**SIMILAR SPECIES**

*P. spadicea* has a similar growth form, but lacks a veil and has pinkish tints in the cap; inedible.

## *AGROCYBE CYLINDRICA*

**A** robust species recognized by the pale cap, ring on the stem, and its densely clustered way of growing.

*Cap* 1 1/4–4 3/4 in/3–12 cm across, convex, cream or pale brown, smooth, somewhat wrinkled and sometimes cracked at the centre. *Gills* at first pale, becoming brown, slightly decurrent or with a decurrent tooth, broad, crowded. *Stem* 2 1/4–5 1/4 × 3/8–3/4 in/6–13 × 1–2 cm, white, often brownish below, fibrillose or becoming smooth, bearing a ring; *ring* membranous, whitish, persistent. *Flesh* mostly white, brown under the cap cuticle and in the stem base, with a slightly mealy smell. *Spore deposit* brown. *Habitat* growing in clusters on old stumps and trunks, especially of poplar, willow and elm. Scarcer in more northerly latitudes. *Edibility* edible.

*AGROCYBE CYLINDRICA*

**SOUTHERN POPLAR MUSHROOM**

**GROUP** Mushrooms and Toadstools

**FAMILY** Bolbitius (Bolbitiaceae)

**SEASON** **EDIBILITY**

**SIMILAR SPECIES**

Unlikely to be confused with other species.

## *KUEHNEROMYCES MUTABILIS*

**A** clustered toadstool recognized by the brown, two-toned cap and dark brown, scaly, ringed stem.

*Cap* 1 1/4–2 1/4 in/3–6 cm across, convex, expanding but often retaining a raised centre, orange-brown or date-brown, soon drying from the centre which becomes pale ochraceous, smooth. *Gills* pale brown, slightly decurrent, crowded. *Stem* 1 1/4–3 × 3/16–3/8 in/3–8 × 0.5–1 cm, pale yellowish and smooth above a thin, brown ring, dark brown and scaly below. *Flesh* whitish or pale yellowish, without a distinctive smell. *Spore deposit* deep ochre. *Habitat* occurring in clusters on stumps and trunks of deciduous trees. *Edibility* edible, but take care to avoid confusion with *Galerina unicolor*.

*KUEHNEROMYCES MUTABILIS*

**TWO-TONE PHOLIOTA or CHANGING PHOLIOTA**

**GROUP** Mushrooms and Toadstools

**FAMILY** Stropharia (Strophariaceae)

**SEASON** **EDIBILITY**

**SIMILAR SPECIES**

The deadly poisonous *Galerina unicolor* (see page 104) has a brownish cap and lacks the scaly stem.

## GALERINA UNICOLOR

**A** common species recognized by the brown colours, clustered habit, stem with ring, and slightly striate cap margin.

*Cap* ¾–2 in/2–5 cm across, at first convex or rather bell-shaped, expanding, smooth, brown, becoming markedly paler, yellowish or pale tan from the centre as it dries, margin long remaining darker, striate. *Gills* adnate, pale brown or ochre, crowded. *Stem* 1½–2¾ × ⅛–3/16 in/4–7 × 0.3–0.5 cm, pale above the ring, dark brown below, the surface bearing whitish fibrils, base often whitish felty; *ring* pale brown, thin, membranous, often lost. *Flesh* pale brown. *Spore deposit* brown. *Habitat* clustered on dead wood and twigs, especially in coniferous woods. *Edibility* **deadly poisonous.**

*GALERINA UNICOLOR*

**MARGINATE GALERINA**

**GROUP** Mushrooms and Toadstools

**FAMILY** Cortinarius (Cortinariaceae)

**SEASON** **EDIBILITY**

**SIMILAR SPECIES**

The edible *Kuehneromyces mutabilis* (see page 103) has a more orange cap and a dark brown, scaly stem. *G. praticola* is more yellowish, smaller, solitary and occurs in grassland; inedible.

## PHOLIOTA SQUARROSA

**F**orms large, dense clusters at the base of tree-trunks, with yellowish brown caps and stems covered with pointed scales.

*Cap* 2¼–4 in/6–10 cm in diameter, convex with a raised centre, ochre-yellow, dry, covered with many shaggy, erect scales in concentric zones. *Gills* adnate, yellowish becoming rusty brown, crowded. *Stem* 2¾–6 × 3/16–⅝ in/7–15 × 0.5–1.5 cm, cylindrical, fibrous, firm, with the colour and scaliness similar to the cap, and a slight ring-zone near to the gill attachment. *Flesh* thick, cream-coloured. *Habitat* dense tufts at the base of deciduous trees, especially ash. *Edibility* poisonous, with a bitter taste, and possibly containing coprine toxins.

*PHOLIOTA SQUARROSA*

**SHAGGY PHOLIOTA**

**GROUP** Mushrooms and Toadstools

**FAMILY** Stropharia (Strophariaceae)

**SEASON** **EDIBILITY**

**SIMILAR SPECIES**

The poisonous Orange Pholiota (*Gymnopilus junonius,* see page 105) grows in similar situations, but lacks the pointed scales on the cap and stem.

## PHOLIOTA AURIVELLA

**A** large, brown-gilled mushroom, growing on the upper branches of living trees, with a slimy scaly, yellow cap and a dry stem.

*Cap* 3–4¾ in/8–12 cm in diameter, bright yellow with a darker centre, slimy, and bearing scattered, flat, slimy, reddish brown scales. *Gills* adnate, yellowish then rusty brown, crowded. *Stem* 3½–6 × ⅜–⅝ in/9–15 × 1–1.5 cm, cylindrical, dry, yellowish brown becoming much darker at the base, bearing a small, fibrous ring towards the apex. *Flesh* thick, pale yellow. *Spore deposit* rusty brown. *Habitat* small clumps on trees, especially beech and maple, often on upper branches. *Edibility* inedible.

*PHOLIOTA AURIVELLA*

**GOLDEN PHOLIOTA**

**GROUP** Mushrooms and Toadstools

**FAMILY** Stropharia (Strophariaceae)

**SEASON** **EDIBILITY**

**SIMILAR SPECIES**

The Shaggy Pholiota (*P. squarrosa,* see this page) has a dry cap, while the Pineapple Pholiota (*P. adiposa*) has a slimy cap and stem; both inedible.

## *HYPHOLOMA FASCICULARE*

**A** common species recognized by the yellow cap and stem, greenish gills and densely clustered habit.

*Cap* ¾–2 in/2–5 cm across, convex, becoming flattened around a raised centre; pale sulphur-yellow, usually more red-brown at the centre; smooth, bearing remnants of the veil at the margin. *Gills* sulphur-yellow when young, becoming olive-green, sinuate, crowded. *Stem* 1½–3 × 3/16–5/16 in/4–8 × 0.5–0.9 cm, often curved, similarly coloured to the cap, smooth or bearing fibrils. *Flesh* thin, yellow. *Spore deposit* purple-brown. *Habitat* dense clusters on and around old stumps and logs of deciduous trees; common. *Edibility* inedible.

*HYPHOLOMA FASCICULARE*

### SULPHUR TUFT

**GROUP** Mushrooms and Toadstools

**FAMILY** Stropharia (Strophariaceae)

**SEASON** **EDIBILITY**

**SIMILAR SPECIES**

*H. sublateritium* has a brick-red cap and lacks olive-green gills. *H. capnoides* also lacks green gills, and occurs on conifer stumps; edible.

## *GYMNOPILUS JUNONIUS*

**O**ften called *Pholiota spectabilis* in older books, this large golden-brown mushroom, with a ring on the stem, forms small clusters at the base of tree-trunks.

*Cap* 2¼–4¾ in/6–12 cm in diameter, convex, tawny brown to golden-brown, dry, radially streaky, with the surface eventually breaking up into small scales. *Gills* adnate, yellowish to rusty brown, and crowded. *Stem* 2¼–6 × ⅜–1¼ in/6–15 × 1–3 cm, robust and cylindrical, fibrous, similarly coloured or paler than the cap, with a large, brown, membranous ring attached near the apex. *Flesh* thick, yellowish. *Spore deposit* rusty brown. *Habitat* dense tufts at base of tree-trunks, especially ash and apple. *Edibility* poisonous.

*GYMNOPILUS JUNONIUS*

### ORANGE PHOLIOTA or BIG LAUGHING GYMNOPILUS

**GROUP** Mushrooms and Toadstools

**FAMILY** Cortinarius (Cortinariaceae)

**SEASON** **EDIBILITY**

**SIMILAR SPECIES**

The inedible Shaggy Pholiota (*Pholiota squarrosa,* see page 104) has erect scales; the edible Honey Fungus (*Armillaria mellea,* page 95) has white, decurrent gills; while the poisonous Jack O'Lantern (*Omphalotus olearius*) lacks a ring on the stem and produces white spores.

## *GYMNOPILUS PENETRANS*

**A** very common toadstool, found in large numbers in conifer woods in the autumn, with a tawny brown cap and rusty brown, spotted gills.

*Cap* 1¼–2¼ in/3–6 cm in diameter, bell-shaped then flattened with a raised centre, tawny yellow, at times fading, smooth, dry, radially streaky. *Gills* 1¼–2¼ × ⅛–¼ in/3–6 × 0.3–0.7 cm, cylindrical, yellowish brown but with a whitish base. *Flesh* thin, yellowish. *Spore deposit* rusty brown. *Habitat* very common in coniferous woods, growing on fallen debris. *Edibility* inedible, with a bitter taste.

*GYMNOPILUS PENETRANS*

### FRECKLE-GILLED GYMNOPILUS

**GROUP** Mushrooms and Toadstools

**FAMILY** Cortinarius (Cortinariaceae)

**SEASON** **EDIBILITY**

**SIMILAR SPECIES**

*G. sapineus* has a more scaly cap, while *G. hybridus* has a cobweb-like veil; both inedible.

## TUBARIA FURFURACEA

**A** very common toadstool, growing on woody debris, and recognized primarily by the brown, decurrent gills, and the white flecks on the cap.

*Cap* $\frac{3}{4}$–$1\frac{1}{2}$ in/2–4 cm in diameter, convex then flattened with a wavy margin, rusty brown when wet but drying much paler from the centre outwards, with tiny, scurfy scales towards the edge. *Gills* short decurrent, cinnamon-brown, and well spaced. *Stem* $\frac{3}{4}$–2 × $\frac{1}{8}$–$\frac{1}{4}$ in/2–5 × 0.3–0.7 cm, similarly coloured to the cap with a woolly base, provided with a faint ring-zone near the gill attachment. *Flesh* thin, brown. *Spore deposit* yellowish brown. *Habitat* found in large numbers, scattered on fallen debris of deciduous trees. *Edibility* inedible.

*TUBARIA FURFURACEA*

**MEALY TUBARIA**

GROUP Mushrooms and Toadstools

FAMILY Cortinarius (Cortinariaceae)

SEASON EDIBILITY

SIMILAR SPECIES

*T. conspersa* is very similar but paler, while *T. hiemalis* is reddish brown and occurs during the winter months; both inedible.

## CREPIDOTUS MOLLIS

**O**ne of the larger slipper toadstools, recognized in the field by an elastic layer which is observed when the cap is pulled apart.

*Cap* $\frac{3}{4}$–$2\frac{3}{4}$ in/2–7 cm in diameter, kidney-shaped to shell-shaped, with a narrow, lateral attachment, pale yellowish brown, drying paler, with a striated margin. *Gills* radiating from a lateral attachment point, white becoming cinnamon-brown, crowded. *Stem* none. *Flesh* thin, white, with a gelatinized layer. *Spore deposit* snuff-brown. *Habitat* on dead and rotting branches, often in large numbers. *Edibility* inedible.

*CREPIDOTUS MOLLIS*

**SOFT SLIPPER TOADSTOOL or JELLY CREPIDOTUS**

GROUP Mushrooms and Toadstools

FAMILY Crepidotus (Crepidotaceae)

SEASON EDIBILITY

SIMILAR SPECIES

*C. calolepsis* has small brown scales on the cap surface; inedible.

## PAXILLUS PANUOIDES

**F**orms tufts of overlapping caps on dead wood; the caps lack a stem and the gills are soft and easily separated from the cap.

*Cap* $1\frac{1}{4}$–4 in/3–10 cm in diameter, spoon-shaped to shell-shaped, dry, greenish yellow with rusty brown spots, and an inrolled margin. *Gills* radiating from a lateral attachment point, yellowish orange, narrow, sometimes forking and interveined, crowded, soft and easily removed by pressing with the thumb. *Stem* absent or nearly so. *Flesh* thin, pale brown. *Spore deposit* yellowish brown. *Habitat* forming clusters on dead wood, also common on domestic wood. *Edibility* edible but poor.

*PAXILLUS PANUOIDES*

**PALE PAXILLUS or STALKLESS PAXILLUS**

GROUP Boletes

FAMILY Paxillus (Paxillaceae)

SEASON EDIBILITY

SIMILAR SPECIES

Most other species of *Paxillus* have a central stem. *Phyllotopsis nidulans* has a hairy cap and yellow gills; inedible. *Crepidotus* species have gills which are not easily separated.

### *RAMARIA STRICTA*

**A** fairly common, yellowish brown coral fungus, found in beech woods, attached to wood by white, mycelial cords.

*Fruitbodies* arise from a root-like base, much branched and coral-like, up to 4 in/10 cm high; individual branches 1/16–3/16 in/0.1–0.5 cm in diameter, erect, yellowish brown to pinkish brown, with the individual branchlets tinted pale yellow at the tips; arising from conspicuous, white, mycelial cords; all parts bruising purplish. *Flesh* thin, white to brownish, tough, with an earthy odour. *Habitat* on or near rotting stumps, or on buried tree-roots, usually of beech, sometimes conifers. *Edibility* edible but with a sour taste, and not recommended.

*RAMARIA STRICTA*

**STRAIGHT-BRANCHED CORAL**

GROUP Club Fungi

FAMILY Coral Fungus (Ramariaceae)

SEASON EDIBILITY

SIMILAR SPECIES

*R. ochraceovirens* has a less erect branching, grows from conifer litter, and stains greenish on bruising; inedible.

### *SPARASSIS CRISPA*

**A** large fungus, resembling a cauliflower, with many short, wavy branches, found growing at the base of pine trees.

*Fruitbody* 4–16 in/10–40 cm in diameter, cauliflower-like, whitish to cream coloured finally brownish, with numerous, flat, ribbon-like branches, 3/8–3/4 in/1–2 cm broad, wavy and leaf-like; arising from a stalk-like base, 3/4–3 × 1 1/4–1 1/2 in/2–8 × 3–4 cm, solid. *Flesh* white, thin, tough, with a sweetish odour. *Spore deposit* cream coloured. *Habitat* at base of conifer-stumps, especially pine. *Edibility* edible and good.

*SPARASSIS CRISPA*

**CAULIFLOWER FUNGUS**

GROUP Club Fungi

FAMILY Cauliflower Fungus (Sparassidaceae)

SEASON EDIBILITY

SIMILAR SPECIES

*S. laminosa* is less branched, with flatter lobes, and usually grows with either beech or oak; edible.

### *AURISCALPIUM VULGARE*

**A**lthough common, this small, dark spine-fungus is difficult to observe, with a stem growing from fallen conifer cones.

*Cap* 3/8–3/4 in/1–2 cm in diameter, kidney-shaped and laterally attached, dark date-brown, coarsely hairy to felt-like. Spines up to 3 mm long and well spaced, hanging vertically on the underside of the cap, pinkish brown. *Stem* 3/4–2 × 1/16–1/8 in/2–5 × 0.2–0.3 cm, lateral, solid, velvety and similarly coloured to the cap. *Flesh* white and tough. *Spore deposit* white. *Habitat* on buried pine, fir or, occasionally, spruce-cones. *Edibility* inedible, with a bitter taste.

*AURISCALPIUM VULGARE*

**FIR-CONE HYDNUM or PINE-CONE TOOTH FUNGUS**

GROUP Tooth Fungi

FAMILY Fir-cone Hydnum (Auriscalpiaceae)

SEASON EDIBILITY

SIMILAR SPECIES

Unlikely to be confused with other species.

## HERICIUM ERINACEUM

A white, fleshy fungus, occasionally found on beech-trunks, with long, pointed spines.

*Fruitbody* 3–6 in/8–15 cm in diameter, cushion-shaped, white, with a poorly developed upper surface but the lower surface densely covered by downwardly pointed, fleshy spines, 1¼–2¾ in/3–7 cm long. *Flesh* white, firm yet sponge-like, succulent. *Spore deposit* white. *Habitat* on trunks of beech or oak trees, rare. *Edibility* rarely eaten in the West, but commercially produced in the Far East.

*HERICIUM ERINACEUM*

**HEDGEHOG FUNGUS or BEARDED TOOTH FUNGUS**

**GROUP** Tooth Fungi

**FAMILY** Hedgehog Fungus (Hericiaceae)

**SEASON** **EDIBILITY**

**SIMILAR SPECIES**

*H. clathroides* is highly branched, while *Creolophus cirrhatus* has smaller, overlapping caps; both edible.

## HERICUM CORALLOIDES

The white fruitbody forms clusters of short branches, each branch bearing downwardly projecting spines along its length.

*Fruitbody* 2–8 in/5–20 cm in diameter, pure white, consisting of a basal trunk from which arise many short, forked branches, 3⁄16–3⁄8 in/0.5–1 cm in diameter, coral-like, each producing fine, pendent spines (3⁄16–3⁄8 in/0.5–1 cm long) along the entire length. *Flesh* white, tough, with an unpleasant odour. *Spore deposit* white. *Habitat* on dead trunks of fir trees, and possibly spruce. *Edibility* inedible.

*HERICIUM CORALLOIDES*

**CORAL SPINE-FUNGUS**

**GROUP** Tooth Fungi

**FAMILY** Hedgehog Fungus (Hericiaceae)

**SEASON** **EDIBILITY**

**SIMILAR SPECIES**

In *H. clathroides* the spines are restricted to the ends of the branches, and the fungus grows on deciduous trees; inedible.

## LENTINUS LEPIDEUS

A tough, whitish toadstool which forms small groups on decayed conifer wood, with a scaly cap and a toothed gill-edge.

*Cap* 1¼–6 in/3–15 cm in diameter, firm, convex to depressed, white to pale yellow, dry, smooth or eventually breaking up into fibrous scales in more or less concentric rings; margin at first inrolled. *Gills* shortly decurrent, whitish, broad, moderately crowded, with a toothed edge. *Stem* ¾–4¼ × ⅜–¾ in/2–11 × 1–2 cm, central or nearly so, cylindrical, solid, similarly coloured to the cap with indistinct scales, and a poorly developed ring near the gill attachment. *Flesh* about ⅜ in/1 cm thick, white, firm. *Spore deposit* white. *Habitat* solitary or in small groups on dead trunks and roots of conifers; often found on domestic timber. *Edibility* edible, but not recommended because too tough.

*LENTINUS LEPIDEUS*

**SCALY LENTINUS**

**GROUP** Mushrooms and Toadstools

**FAMILY** Lentinus and Oyster Cap (Lentinaceae)

**SEASON** **EDIBILITY**

**SIMILAR SPECIES**

The Large Lentinus (*L. ponderosus*), which grows in western North America, is more fleshy and lacks a ring on the stem; inedible.

## LENTINUS TIGRINUS

**A** tough, pale toadstool, with numerous small, dark scales, growing on dead wood mostly of poplar and willow.

*Cap* 1/16–4 in/1–10 cm in diameter, tough, convex with a depressed centre to funnel-shaped, at first greyish brown with radially arranged, small, fibrous scales, becoming whitish on expansion. *Gills* crowded, narrow, whitish, with a toothed edge. *Stem* ¾–4 × 3/16–3/8 in/2–10 × 0.5–1 cm, central or nearly so, solid, yellowish, covered with small grey-brown scales; with a cobweb-like annular zone below the gill attachment. *Flesh* 1/16–1/8 in/0.2–0.3 cm thick, firm, white, fibrous. *Spore deposit* white. *Habitat* forming clusters on old branches, especially poplar and willow, in damp places. *Edibility* inedible, too tough.

*LENTINUS TIGRINUS*

### TIGER-SPOT LENTINUS

**GROUP** Mushrooms and Toadstools

**FAMILY** Lentinus and Oyster Cap (Lentinaceae)

**SEASON** **EDIBILITY**

**SIMILAR SPECIES**

In North America there is a form called *Lentodium squamulosum,* in which the gills are deformed and permanently covered by a membranous veil. *Lentinus strigosus* is covered with coarse hairs, and the gill-edge is not toothed. Both inedible.

## POLYPORUS SQUAMOSUS

**T**his polypore forms large, plate-like, overlapping, yellowish caps covered with zones of blackish brown scales.

*Cap* 4–20 in/10–50 cm in diameter, laterally to almost centrally attached to the stem, kidney-shaped to semi-circular, yellowish brown with blackish brown, pointed scales arranged in concentric zones, and a thin, downturned margin. *Tubes* decurrent, up to 5/16 in/0.7 cm long; *pores* large and angular, 1–2 per 1/16 in/0.1 cm, cream-coloured, and radially elongated. *Stem* 1¼–4 × 3/8–2 in/3–10 × 1–5 cm, cylindrical or tapered towards the base, solid, pale above, with a black velvety base. *Flesh* ¾–1½ in/2–4 cm thick, white, with a sweetish odour and a floury taste. *Spore deposit* white. *Habitat* on trunks of deciduous trees, particularly beech and elm, also maple and walnut, usually as a wound parasite. *Edibility* said to be edible when very young but not recommended, soon attacked by insects.

*POLYPORUS SQUAMOSUS*

### DRYAD'S SADDLE

**GROUP** Bracket-fungi

**FAMILY** True Polyporus (Polyporaceae)

**SEASON** **EDIBILITY**

**SIMILAR SPECIES**

*P. tuberaster* is smaller, usually has a central stem, and grows from an underground tuber (sclerotium); inedible.

## POLYPORUS VARIUS

**A** fairly small, funnel-shaped polypore, with decurrent, greyish pores and a black stem base.

*Cap* 2–3 in/5–8 cm in diameter, fan-shaped to funnel-shaped and depressed, pale orange-brown, smooth or wrinkled, with a thin, wavy margin. *Tubes* decurrent, about 1/8 in/0.3 cm long; *pores* per 1/16 in/0.1 cm, yellowish brown to greyish brown. *Stem* 3/8–2¼ × 3/16–5/8 in/1–6 × 0.5–1.5 cm, central or laterally attached, pale above but the lower region black and wrinkly. *Flesh* about 1/8 in/0.3 cm thick, white, firm, with a pleasant odour. *Spore deposit* white. *Habitat* on stumps and trunks of deciduous trees, especially beech and alder. *Edibility* inedible because too tough.

*POLYPORUS VARIUS*

### ELEGANT POLYPORE

**GROUP** Bracket-fungi

**FAMILY** True Polyporus (Polyporaceae)

**SEASON** **EDIBILITY**

**SIMILAR SPECIES**

*P. nummularius* is a small species, and *P. melanopus* has a brown to purplish cap, white pores and a stem which is entirely black; *P. badius* has a shiny cap and a black stem, and *P. brumalis* lacks a black crust on the stem. All these specimens are inedible.

## CORIOLUS VERSICOLOR

An extremely common bracket-fungus, forming clusters and overlapping tiers on dead wood, recognized by the thin, tough caps with a strongly zoned surface, and pale flesh and pores.

*Cap* 1¼–3 in/3–8 cm in diameter, thin, fan-shaped, with the surface varying from grey to reddish brown or even black, with many narrow zones, which are at first velvety but gradually become smooth on weathering. *Tubes* shallow, about 1/16 in/0.1–0.2 cm long, white; *pores* small, 3–5 per 1/16 in/0.1 cm, angular. *Flesh* about 1/16 in/0.2 cm thick, white, tough. *Spore deposit* white. *Habitat* groups on dead deciduous trees and stumps, sometimes on conifers. *Edibility* inedible.

| *CORIOLUS VERSICOLOR* |
|---|
| **VARICOLOURED BRACKET or TURKEY-TAIL** |
| **GROUP** Bracket-fungi |
| **FAMILY** Poroid Bracket (Coriolaceae) |
| **SEASON** · **EDIBILITY** |
| **SIMILAR SPECIES** *C. hirsutus* is grey and densely hairy, and usually grows on beech; *C. pubescens* is straw-coloured, with a soft-hairy surface; both inedible. |

## BJERKANDERA ADUSTA

A small bracket-fungus forming tiers on dead wood, recognized by the dark grey tube-layer which contrasts with the white flesh when broken open.

*Cap* ¾–2¾ in/2–7 cm in diameter, convex, fan-shaped with a broad attachment to the bark, pale grey, often with a blackish region towards the margin, smooth and faintly zoned. *Tubes* decurrent, dark grey, up to 2 mm long; *pores* small, 4–6 per 1/16 in/0.1 cm, grey to blackish. *Flesh* up to 5 mm thick, whitish, with a characteristic black line immediately above the tubes. *Spore deposit* white. *Habitat* clustered on dead wood of deciduous trees, especially beech. *Edibility* inedible.

| *BJERKANDERA ADUSTA* |
|---|
| **SMOKY POLYPORE** |
| **GROUP** Bracket-fungi |
| **FAMILY** Poroid Bracket (Coriolaceae) |
| **SEASON** · **EDIBILITY** |
| **SIMILAR SPECIES** *B. fumosa* is generally thicker and has a paler pore surface; inedible. |

## HIRSCHIOPORUS ABIETINUS

Forms thin brackets on coniferous tree-trunks, and recognized by the violaceous tints on the pore surface.

*Cap* ¾–1½ in/2–4 cm in diameter, shelf-like with a broad attachment to the bark, pale greyish brown, with concentric ridges, velvety to hairy, with a wavy, paler margin. *Tubes* 1/16–⅛ in/0.2–0.3 cm long; *pores* small, 4–5 per 1/16 in/0.1 cm, bright violet-coloured when young but gradually fading to become greyish pink. *Flesh* thin, tough, brown. *Spore deposit* white. *Habitat* forming many tiers on trunks of coniferous trees, including fir, pine and spruce. *Edibility* inedible.

| *HIRSCHIOPORUS ABIETINUS* |
|---|
| **FIR POLYSTICTUS** |
| **GROUP** Bracket-fungi |
| **FAMILY** Poroid Bracket (Coriolaceae) |
| **SEASON** · **EDIBILITY** |
| **SIMILAR SPECIES** *H. fuscoviolaceus* prefers pine, and has toothed-spiny pore surfaces; inedible. |

## *HAPALOPILUS NIDULANS*

**A** small bracket-fungus distinguished by the cinnamon-brown colour, and the unique purplish colour change when alkaline solution is applied to the surface.

*Cap* 3/4–3 in/2–8 cm in diameter, shelf-like with a broad attachment, slightly convex, cinnamon-brown, smooth, with a thin, sharp edge. *Tubes* 3/16–3/8 in/0.4–1 cm long, greyish brown; *pores* 2–4 per 1/16 in/0.1 cm, cinnamon-brown, angular. *Flesh* up to 3/4 in/2 cm thick, cinnamon-brown, tough. *Spore deposit* white. *Habitat* on dead wood of beech, birch and fir; also occurring on pine and spruce. *Edibility* inedible.

*HAPALOPILUS NIDULANS*

**TENDER NESTING POLYPORE**

**GROUP** Bracket-fungi

**FAMILY** Poroid Bracket (Coriolaceae)

**SEASON** **EDIBILITY**

**SIMILAR SPECIES**

*Aurantioporus croceus* is a very rare, orange species, growing on oak, and stains red with alkaline solution; inedible.

## *DAEDALEA QUERCINA*

**A** large, pale brown bracket, growing on oak-stumps, characterized by a pore surface which forms a maze pattern.

*Cap* 4–6 in/10–15 cm in diameter, shelf-like, broadly attached, with a flattened upper surface, yellowish brown to greyish brown, with a rigid, narrow edge. *Tubes* up to 1 1/2 in/4 cm long, yellowish; *pores* large and very irregular, partially fusing with each other to form a maze-like pattern. *Flesh* about 3/8 in/1 cm thick, yellowish brown, tough and corky. *Spore deposit* white. *Habitat* forms one to several brackets on oak-stumps or, occasionally, on chestnut. *Edibility* inedible.

*DAEDALEA QUERCINA*

**MAZEGILL or THICK-MAZE OAK POLYPORE**

**GROUP** Bracket-fungi

**FAMILY** Poroid Bracket (Coriolaceae)

**SEASON** **EDIBILITY**

**SIMILAR SPECIES**

*Gloeophyllum* species have a smaller habit, and a thin, dark flesh.

## *DAEDALEOPSIS CONFRAGOSA*

**C**ommonly found bracket-fungi, with a maze-patterned pore surface, which bruises pinkish brown.

*Cap* 2–6 in/5–15 cm in diameter, shelf-shaped with broad attachment, a convex to flattened upper surface, at first whitish then darkening to brown or purplish brown, with concentric zones and radial streaks. *Tubes* about 3/8 in/1 cm long, whitish to grey-brown; *pores* large, 1–2 per 1/16 in/0.1 cm, sometimes as round pores but often radially elongated or forming a maze pattern, white, bruising pinkish brown. *Flesh* about 3/16 in/0.5 cm thick, yellowish brown, tough and corky. *Spore deposit* white. *Habitat* on deciduous trees, especially alder, beech and willow. *Edibility* inedible.

*DAEDALEOPSIS CONFRAGOSA*

**BLOOD-STAINED BRACKET or THIN-MAZE FLAT POLYPORE**

**GROUP** Bracket-fungi

**FAMILY** Poroid Bracket (Coriolaceae)

**SEASON** **EDIBILITY**

**SIMILAR SPECIES**

The variety *tricolor* is thinner, and blackish red with narrow zoning; inedible.

## *FOMES FOMENTARIUS*

One of the largest of the bracket-fungi, found on either beech or birch, forming hoof-shaped fruitbodies with a pale brown, corky flesh.

*Cap* 4–20 in/10–50 cm in diameter, very thick, horse hoof-shaped and layered, with a pale grey to blackish surface, smooth, and a broadly rounded margin. *Tubes* forming annual layers, each layer about ⅜ in/1 cm long; *pores* 3–4 per 1/16 in/0.1 cm, round, yellowish brown, bruising darker. *Flesh* up to 1¼ in/3 cm thick, pale brown, very tough and leathery. *Spore deposit* white. *Habitat* on birch, particularly in northern areas, and beech. *Edibility* inedible.

*FOMES FOMENTARIUS*

**TINDER FUNGUS**

**GROUP** Bracket-fungi

**FAMILY** Poroid Bracket (Coriolaceae)

**SEASON** **EDIBILITY**

**SIMILAR SPECIES**

*Phellinus igniarius* has a dark brown flesh, while *Fomitopsis pinicola* (see this page) has a whitish pore surface, and the large *Ganoderma* species produce a brown spore deposit; all inedible.

## *FOMITOPSIS PINICOLA*

A common bracket-fungus of spruce trees, distinguished by the resinous crust on the surface and the white flesh.

*Cap* 2–9 in/5–25 cm in diameter, very thick, hoof-shaped and layered with a thin, brittle, resinous crust, at first shiny and yellow or reddish orange, finally black and dull, with a rounded, whitish margin. *Tubes* layered, each layer about ⅜ in/1 cm long; *pores* 3–4 per 1/16 in/0.1 cm, whitish, bruising grey. *Flesh* up to 1½ in/4 cm thick, whitish to cream-ochre, tough and corky, with an acidic odour when young. *Spore deposit* white. *Habitat* common, typically on spruce trees, but other hosts. *Edibility* inedible.

*FOMITOPSIS PINICOLA*

**RED-RIMMED BRACKET**

**GROUP** Bracket-fungi

**FAMILY** Poroid Bracket (Coriolaceae)

**SEASON** **EDIBILITY**

**SIMILAR SPECIES**

The cream-coloured flesh distinguishes this from other *Phellinus* species. *Fomes fomentarius* (see this page) lacks the resinous crust. All these species are tough and inedible.

## *HETEROBASIDION ANNOSUM*

The small irregularly shaped brackets of this serious rot of spruce trees are found at the base of living trees.

*Cap* forming irregular, indefinite lobes, up to 8 in/20 cm long and 1½ in/4 cm thick, with a pale to dark brown, dull surface which is wrinkled by many zones and forms a crust. *Tubes* layered, each layer about 3/16 in/0.4 cm long; *pores* 2–5 per 1/16 in/0.1 cm, whitish, forming an uneven surface. *Flesh* almost ⅜ in/1 cm thick, white, hard to corky. *Spore deposit* white. *Habitat* on and around the roots and base of conifer-trunks, especially spruce. *Edibility* inedible.

*HETEROBASIDION ANNOSUM*

**ROOT FOMES or CONIFER-BASE POLYPORE**

**GROUP** Bracket-fungi

**FAMILY** Poroid Bracket (Coriolaceae)

**SEASON** **EDIBILITY**

**SIMILAR SPECIES**

The irregular shelf-like brackets and habit at base of spruce trees should distinguish this species from other polypores.

## *LAETIPORUS SULPHUREUS*

The fruitbody forms large clusters of overlapping, sulphur-yellow caps, growing on the trunks of both living and dead trees.

*Cap* 2–8 in/5–20 cm in diameter, fan-shaped, flattened but up to 2 in/5 cm thick, with an orange, lemon-yellow or sulphur-yellow surface becoming paler when old except for the margin, smooth. *Tubes* forming a narrow layer, about 3/16 in/0.4 cm deep; *pores* small, 3–5 per 1/16 in/0.1 cm, round, sulphur-yellow. *Flesh* thick, pale yellowish cream to almost white, at first soft and moist but becoming crumbly like chalk. *Spore deposit* white. *Habitat* clustered on living trees, generally preferring oak and yew. *Edibility* edible.

*LAETIPORUS SULPHUREUS*

**SULPHUR POLYPORE or CHICKEN OF THE WOODS**

GROUP Bracket-fungi

FAMILY Poroid Bracket (Coriolaceae)

SEASON EDIBILITY

SIMILAR SPECIES

The distinctive yellow brackets distinguish this species from other polypores.

## *MERIPILUS GIGANTEUS*

A compound fruitbody, up to 24 in/60 cm across, formed of many caps, growing from the roots of beech and oak trees. A short-lived species, it soon rots.

*Cap* 2–8 in/5–20 cm in diameter, laterally attached, fan-shaped with a tapering base, yellowish brown, with concentric, darker zones, smooth, with a thin margin. *Tubes* decurrent, up to 3/8 in/1 cm deep, often much less; *pores* 3–5 per 1/16 in/0.1 cm, pale cream, bruising blackish. *Flesh* up to 3/4 in/2 cm thick, fibrous, white but becoming blackened when broken open. *Spore deposit* white. *Habitat* forming large tufts and rosettes on roots and at base of stumps, especially beech and oak. *Edibility* inedible.

*MERIPILUS GIGANTEUS*

**GIANT POLYPORE or BLACK-STAINING POLYPORE**

GROUP Bracket-fungi

FAMILY Poroid Bracket (Coriolaceae)

SEASON EDIBILITY

SIMILAR SPECIES

*Grifola frondosa* (see this page) forms similar tufts but has smaller caps and does not bruise black. *Bondarzewia berkeleyi*, in North America, is cream-coloured and also does not bruise black. Both species are inedible.

## *GRIFOLA FRONDOSA*

Resembling a cauliflower, up to 12 in/30 cm across, with numerous, greyish brown caps, arising from a short, stout stem, attached to buried tree roots.

*Cap* 3/4–2 in/2–5 cm in diameter, fan-shaped and lobed, greyish brown, with radial streaks and a thin, wavy edge. *Tubes* decurrent, about 1/8 in/0.3 cm long; *pores* 2–3 per 1/16 in/0.1 cm, white. *Stem* short and stout, solid, whitish. *Flesh* 1/8–3/16 in/0.3–0.5 cm thick, white, fibrous, with either a sweetish or developing a very unpleasant odour. *Spore deposit* white. *Habitat* tufted, growing from buried roots of deciduous trees, especially ash and oak. *Edibility* edible when young.

*GRIFOLA FRONDOSA*

**HEN OF THE WOODS**

GROUP Bracket-fungi

FAMILY Poroid Bracket (Coriolaceae)

SEASON EDIBILITY

SIMILAR SPECIES

The Umbrella Polypore (*Polyporus umbellatus*, page 64) is similar but has circular caps, each with a central stipe, and the entire fruitbody arises from an underground tuber (sclerotium); inedible.

## *PIPTOPORUS BETULINUS*

**A**lways found in large numbers on birch trees, and easily recognized by the smooth, rounded caps.

*Cap* 2–8 in/5–20 cm in diameter, round, hoof-shaped with a narrow, lateral attachment to the host; surface at first white soon becoming greyish brown, not zoned, smooth, with a broadly rounded edge and an inrolled margin. *Tubes* up to ⅜ in/1 cm long, white, developing late; *pores* small, 3–4 per 1/16 in/0.1 cm, whitish discolouring brownish. *Flesh* up to 1¼ in/3 cm thick, pure white and corky. *Spore deposit* white. *Habitat* on trunks of living and dead birch trees. *Edibility* inedible.

*PIPTOPORUS BETULINUS*

**RAZOR-STROP FUNGUS or BIRCH POLYPORE**

GROUP Bracket-fungi

FAMILY Poroid Bracket (Coriolaceae)

SEASON — EDIBILITY

SIMILAR SPECIES

Always confined to birch trees, this species is unlikely to be confused. *Lenzites betulina* (see this page) has radiating gills instead of pores on the underside; inedible.

## *LENZITES BETULINA*

**A** thin but tough bracket-fungus, with radiating gills on the underside instead of tubes and pores.

*Cap* ¾–3½ in/2–9 cm in diameter, bracket to shelf-shaped with a broad or narrow, lateral attachment, pale greyish brown (or discoloured greenish by algae), finely hairy, with narrow, concentric zoning. *Pore surface* with thick, radiating gills, unequal and forked, 12 to 15 per ⅜ in/1 cm, cream coloured to greyish brown. *Flesh* 1/16–⅛ in/0.2–0.3 cm thick, white, tough. *Spore deposit* white. *Habitat* on dead wood of birch, but also found on beech and oak, never common. *Edibility* inedible.

*LENZITES BETULINA*

**BIRCH LENZITES**

GROUP Bracket-fungi

FAMILY Poroid Bracket (Coriolaceae)

SEASON — EDIBILITY

SIMILAR SPECIES

The upper surface can easily be confused with *Coriolus* species but the gills on the underside are characteristic.

## *GLOEOPHYLLUM SEPIARIUM*

**S**mall brackets forming tiers on dead spruce wood, with a bright yellow margin.

*Cap* 1¼–4 in/3–10 cm in diameter, and up to 2¾ in/7 cm wide, semi-circular with a broad lateral attachment, yellowish brown to reddish brown or almost black; with a conspicuous yellow margin. *Pore surface* gill-like with radially elongated pores, 1–2 per 1/16 in/0.1 cm, golden brown or darker. *Flesh* about 3/16 in/0.5 cm thick, dark brown. *Spore deposit* white. *Habitat* common, on spruce-stumps and logs, rarely found on deciduous trees, but often on domestic wood. *Edibility* inedible.

*GLOEOPHYLLUM SEPIARIUM*

**YELLOW-RED GILL POLYPORE**

GROUP Bracket-fungi

FAMILY Poroid Bracket (Coriolaceae)

SEASON — EDIBILITY

SIMILAR SPECIES

The rare *G. abietinum* grows on the same host but differs in having much more widely spaced gills; inedible.

## *TYROMYCES CAESIUS*

A small, soft-fleshed, white bracket-fungus, which is easily recognized by the bluish stains.

*Cap* $\frac{3}{8}$–$1\frac{1}{2}$ in/1–4 cm in diameter, semicircular with a broad, lateral attachment, velvety to hairy, white but bruising bluish grey, not zoned. *Tubes* about $\frac{3}{16}$–$\frac{5}{16}$ in/0.5–0.8 cm long; *pores* 4–5 per $\frac{1}{16}$ in/0.1 cm, soon bruising bluish grey. *Flesh* white, soft, fibrous, with a mild taste and a fruity odour. *Spore deposit* white. *Habitat* on dead wood of conifers. *Edibility* inedible.

*TYROMYCES CAESIUS*

**BLUE-CHEESE POLYPORE**

**GROUP** Bracket-fungi

**FAMILY** Poroid Bracket (Coriolaceae)

**SEASON** **EDIBILITY**

**SIMILAR SPECIES**

*T. subcaesius* grows on deciduous trees, while *T. stipticus* has a roughened upper surface, grows on conifers, and has a very bitter taste; both inedible.

## *TYROMYCES LACTEUS*

A soft, white bracket, commonly found on deciduous trees.

*Cap* $\frac{3}{4}$–$2\frac{1}{4}$ in/2–6 cm in diameter, semicircular with a broad lateral attachment, rather thick at first, developing greyish tints, smooth or roughened. *Tubes* $\frac{3}{8}$–$\frac{5}{8}$ in/1–1.5 cm long, white: *pores* 3–5 per $\frac{1}{16}$ in/0.1 cm, white. *Flesh* up to $\frac{1}{8}$ in/0.3 cm thick, soft, white, fibrous. *Spore deposit* white. *Habitat* on deciduous trees. *Edibility* inedible.

*TYROMYCES LACTEUS*

**MILK-WHITE POLYPORE**

**GROUP** Bracket-fungi

**FAMILY** Poroid Bracket (Coriolaceae)

**SEASON** **EDIBILITY**

**SIMILAR SPECIES**

The inedible White Cheese Polypore (*T. chioneus*) has a fragrant smell, and dries yellowish.

## *PYCNOPORUS CINNABARINUS*

An easily recognized, small bracket fungus, with all parts coloured bright orange-red.

*Cap* $\frac{3}{4}$–4 in/2–10 cm in diameter, semicircular with a broad lateral attachment, orange-red discolouring paler in old specimens, smooth or uneven, with a thin, straight margin. *Tubes* $\frac{3}{16}$–$\frac{1}{4}$ in/0.4–0.6 cm long; *pores* 2–3 per $\frac{1}{16}$ in/0.1 cm, deep orange-red. *Flesh* up to $\frac{3}{4}$ in/2 cm thick, tough and corky, orange-red, lacking an odour. *Spore deposit* white. *Habitat* on dead wood of both deciduous trees, especially oak, and conifers; sometimes several caps are joined by their sides. *Edibility* inedible.

*PYCNOPORUS CINNABARINUS*

**CINNABAR-RED POLYPORE**

**GROUP** Bracket-fungi

**FAMILY** Poroid Bracket (Coriolaceae)

**SEASON** **EDIBILITY**

**SIMILAR SPECIES**

The orange-red is so distinctive that it cannot be confused. In southern North America, *P. sanguineus* is found which is usually much thinner; inedible.

## GANODERMA APPLANATUM

**A** large, perennial polypore growing solitary or in groups on dead stumps, recognized by the flat caps with a hard upper crust, and a white pore surface.

*Cap* 4–20 in/10–50 cm in diameter, plate-like, with the upper surface forming a dull, hard crust with indistinct concentric zones, at first pale brown then dark grey-brown (although often covered with a brown spore deposit), and a thin, sharp margin. *Tubes* layered, each layer 3/16–3/8 in/0.5–1 cm long, brown; *pores* 5–6 per 1/16 in/0.1 cm, whitish becoming brownish. *Flesh* woody brown, often white flecks, corky, thinner than the tube-layer. *Spore deposit* cinnamon-brown. *Habitat* on dead wood of beech, maple, ash, poplar and oak, rarely on conifers. *Edibility* inedible.

*GANODERMA APPLANATUM*

**ARTIST'S FUNGUS**

**GROUP** Bracket-fungi

**FAMILY** Ganoderma (Ganodermataceae)

**SEASON** **EDIBILITY**

**SIMILAR SPECIES**

*G. adspersum* is often confused yet probably more common, at least in western Europe, differing in the dark brown flesh, lacking white flecks, which is thicker than the tube-layer, and the round edge to the cap; inedible.

## GANODERMA LUCIDUM

**A** stalked species of *Ganoderma,* easily recognized by the strongly varnished crust of the cap and stem.

*Cap* 1¼–12 in/3–30 cm in diameter, kidney-shaped and semi-circular, attached to a lateral stem, dark red to blackish with a white to yellowish brown margin, covered by a highly polished varnish. *Tubes* up to ¾ in/2 cm long, brown; *pores* 3–4 per 1/16 in/0.1 cm, greyish, staining brown. *Stem* 2–6 × 3/8–1¼ in/5–15 × 1–3 cm, cylindrical or flattened, smooth, dark red to black, with a varnished crust. *Flesh* pale brown, corky. *Spore deposit* cinnamon-brown. *Habitat* on roots of oak, also other deciduous trees. *Edibility* inedible.

*GANODERMA LUCIDUM*

**SHINING GANODERMA**

**GROUP** Bracket-fungi

**FAMILY** Ganoderma (Ganodermataceae)

**SEASON** **EDIBILITY**

**SIMILAR SPECIES**

*G. carnosus* in Europe is very similar but grows on stumps of fir trees; inedible. In North America, *G. curtisii* has a bright ochre-brown cap; inedible. The Chinese "Mushroom of Immortality" or Ling Chih is closely related.

## GANODERMA TSUGAE

**A** large *Ganoderma* species, with a shiny, varnished surface, distinguished by a soft, white flesh.

*Cap* 2–12 in/5–30 cm in diameter, kidney-shaped or fan-shaped, reddish brown to black, highly varnished, smooth, with some concentric zoning, and often with a white margin. *Tubes* up to 3/8 in/1 cm long, brown; *pores* 4–6 per 1/16 in/0.1 cm, whitish becoming brown. *Stem* mostly present and laterally attached, up to 6 in/15 cm long, 3/8–1½ in/1–4 cm thick, with a varnished crust similar to the cap. *Flesh* up to 1¼ in/3 cm thick, white, soft and fleshy in young specimens, becoming firm. *Spore deposit* cinnamon-brown. *Habitat* on dead wood of coniferous trees, only found in North America. *Edibility* inedible.

*GANODERMA TSUGAE*

**HEMLOCK VARNISH GANODERMA**

**GROUP** Bracket-fungi

**FAMILY** Ganoderma (Ganodermataceae)

**SEASON** **EDIBILITY**

**SIMILAR SPECIES**

*G. oregonense* is larger and dark coloured; inedible.

## *FISTULINA HEPATICA*

**A** large, reddish, fleshy bracket-fungus, growing on oak trunks, and characterized by the tubes on the lower surface which are not fused to each other.

*Cap* 2¾–8 in/7–20 cm in diameter, tongue-like or bracket-shaped, convex and attached by a short, lateral stalk, pinkish to orange-red, with a sticky, roughened surface. *Tubes* up to ⅜ in/1 cm long, not fused but each developing separately; *pores* 2–3 per 1/16 in/0.1 cm, whitish or yellowish, bruising reddish brown. *Flesh* ¾–2 in/2–5 cm thick, succulent, whitish soon streaked with red. *Spore deposit* white to pale pink. *Habitat* on trunks of living or dead oak trees, sometimes chestnut. *Edibility* edible but somewhat acidic.

*FISTULINA HEPATICA*

**BEEFSTEAK FUNGUS**

**GROUP** Bracket-fungi

**FAMILY** Beefsteak Fungus (Fistulinaceae)

**SEASON** **EDIBILITY**

**SIMILAR SPECIES**

The individually separated tubes distinguish this from all other pore-fungi.

## *PHAEOLUS SCHWEINITZII*

**This** centrally stalked, bracket-fungus forms large, plate-like caps with a sulphur-yellow margin and a brown flesh; it is found at the base of living or dead conifer trees.

*Cap* 3–12 in/8–30 cm in diameter, flat and plate-like, with concentric orange zones and a dark reddish brown centre, felty, with a sulphur-yellow margin; entire fruit-body becoming blackish on weathering. *Tubes* up to ⅜ in/1 cm long, brown, brittle; *pores* 2–5 per 1/16 in/0.1 cm, often splitting and maze-like, greenish yellow discolouring brown. *Stem* 1¼–3 × ¾–2 in/3–8 × 2–5 cm, short, solid, dark rusty brown. *Flesh* ¾–1¼ in/2–3 cm thick, brown, soft at first becoming fibrous and brittle. *Spore deposit* white. *Habitat* on dead or living wood at the base of pine, larch or spruce, sometimes forming small clusters. *Edibility* inedible.

*PHAEOLUS SCHWEINITZII*

**LARGE PINE POLYPORUS or DYE POLYPORE**

**GROUP** Bracket-fungi

**FAMILY** Brown Polypore (Hymenochaetaceae)

**SEASON** **EDIBILITY**

**SIMILAR SPECIES**

*Coltricia* species have a central stem and brown flesh, but are much smaller and thinner; inedible.

## *INONOTUS DRYADEUS*

**A** broadly attached polypore, found growing at the base of oak-trunks, and easily recognized by the reddish brown flesh and by the exudation of watery droplets at the cap margin.

*Cap* 4–9 in/10–25 cm in diameter, strongly convex and broadly attached, thick, with an uneven, pitted upper surface, at first cream-coloured later orange-brown, oozing reddish, watery droplets over the rounded margin. *Tubes* 3/16–¾ in/0.5–2 cm long, brown; *pores* small, 3–4 per 1/16 in/0.1 cm, whitish to yellowish brown. *Flesh* reddish brown, succulent-tough, with an unpleasant, acidic odour. *Spore deposit* white. *Habitat* at the base of living oak-trunks, occasionally chestnut. *Edibility* inedible.

*INONOTUS DRYADEUS*

**WEEPING POLYPORE or WARTED OAK POLYPORE**

**GROUP** Bracket-fungi

**FAMILY** Brown Polypore (Hymenochaetaceae)

**SEASON** **EDIBILITY**

**SIMILAR SPECIES**

Apart from in Britain, *I. dryophilus* also occurs on oak, but differs in having a large, granular core at the base of the flesh, and a brown spore deposit; inedible.

## INONOTUS RADIATUS

**A** small, brown-fleshed polypore, usually forming several small brackets on dead alder branches.

*Cap* 1¼–3 in/3–8 cm in diameter, bracket-like and broadly attached, with a finely velvety upper surface which is radially wrinkled and concentrically zoned, reddish brown with a thin, yellow margin. *Tubes* up to ⅜ in/1 cm long, brown; *pores* small, 2–5 per 1/16 in/0.1 cm, whitish to greyish brown, with a greyish iridescence. *Flesh* about 3/16 in/0.5 cm thick, rusty brown, succulent-fibrous, brittle when dry. *Spore deposit* cream coloured. *Habitat* usually in tiers, on trunks and branches of alder trees. *Edibility* inedible.

*INONOTUS RADIATUS*

### ALDERWOOD POLYPORE

GROUP Bracket-fungi

FAMILY Brown Polypore (Hymenochaetaceae)

SEASON EDIBILITY

SIMILAR SPECIES

The Mustard Yellow Polypore (*Phellinus gilvus*) is more brightly coloured, thinner, and grows on a range of tree species; inedible.

## PHELLINUS IGNIARIUS

**A** hard bracket-fungus, with a smooth crust on the upper surface, characterized by the dark brown flesh and growing on trunks of apple or willow trees.

*Cap* 2–9 in/5–25 cm in diameter, cushion-like, with a broad, lateral attachment, up to 6 in/15 cm thick at the base, with a dull, pale to blackish grey, smooth upper surface, broadly furrowed and with a rounded, pale margin. *Tubes* layered, each layer up to 3/16 in/0.5 cm deep, brown; *pores* small, 5–6 per 1/16 in/0.1 cm, reddish brown becoming greyish. *Flesh* dark orange-brown, hard and tough. *Spore deposit* white. *Habitat* generally solitary, on trunks of living apple trees, at times also on willow and some other deciduous trees. *Edibility* inedible.

*PHELLINUS IGNIARIUS*

### FALSE TINDER FUNGUS or FLECK-FLESH POLYPORE

GROUP Bracket-fungi

FAMILY Brown Polypore (Hymenochaetaceae)

SEASON EDIBILITY

SIMILAR SPECIES

*P. robustus* has a yellowish brown flesh and grows on oak, *P. nigricans* has a black crust and grows on birch, and *P. tremulae* is found on poplar trees. All three specimens are inedible.

## PHLEBIA RADIATA

**A** pinkish orange crust, with radial furrows and warty surface, found on the bark of deciduous trees.

*Fruitbody* forms a flattened crust on bark, only occasionally with a free margin, at first consisting of small round patches but expanding to 1¼–1½ in/3–4 cm in diameter, and several patches often joining, pale orange to violet-grey, radially furrowed and very warty over the surface, paler at the margin. *Flesh* thin, somewhat gelatinous but drying horny. *Habitat* on bark of dead wood of deciduous trees, rarely conifers. *Edibility* inedible.

*PHLEBIA RADIATA*

### VEIN CRUST or RADIATING PHLEBIA

GROUP Bracket-fungi

FAMILY Merulius (Meruliaceae)

SEASON EDIBILITY

SIMILAR SPECIES

*P. rufa* is very similar but lacks the radial furrows; inedible.

## *MERULIUS TREMELLOSUS*

**S**mall gelatinous brackets, with a pinkish orange, wrinkled lower surface, found growing on dead wood.

*Fruitbody* spreads over the wood and curves at the upper margin to form small brackets, ¾–1½ in/2–4 cm wide. *Upper surface* white to pale yellowish pink, hairy or felty, with a thin, wavy margin. *Lower surface* radially wrinkled and folded ("merulioid") to almost form pores, yellowish orange to salmon-pink. *Flesh* thin, soft and gelatinous, drying horny. *Spore deposit* white. *Habitat* on rotting wood, especially growing on the underside of fallen trunks. *Edibility* worthless.

*MERULIUS TREMELLOSUS*

**JELLY ROT or TREMBLING MERULIUS**

**GROUP** Bracket-fungi

**FAMILY** Merulius (Meruliaceae)

**SEASON** **EDIBILITY**

**SIMILAR SPECIES**

This common, gelatinous bracket-fungus is unlikely to be confused with other species.

## *SERPULA LACRYMANS*

**T**he well-known, destructive fungus of domestic wood, characterized by fruitbodies with a rusty brown, strongly wrinkled, lower surface; spreading by greyish brown, fibrous cords.

*Fruitbody* spreading over the substrate and forming large, pancake-like patches, up to ⅜ in/1 cm thick, eventually the margin curving to form indefinite brackets. *Upper surface* white to cream-coloured, bruising brownish, with a broad, white margin. *Lower surface* strongly wrinkled and folded to irregularly pore-like, olive-brown to deep rusty brown. *Flesh* white, tough, fibrous, with a pungent odour. *Spore deposit* yellowish brown. *Habitat* on domestic conifer wood found in damp, poorly ventilated buildings. *Edibility* inedible.

*SERPULA LACRYMANS*

**DRY ROT FUNGUS**

**GROUP** Bracket-fungi

**FAMILY** Dry and Wet Rot (Coniophoraceae)

**SEASON** **EDIBILITY**

**SIMILAR SPECIES**

*S. himantioides* grows on forest conifers and is much thinner in appearance; inedible.

## *CONIOPHORA PUTEANA*

**A** fungus commonly found in damp buildings on conifer wood, and distinguished by the olive-brown, warty surface and a whitish, cottony margin.

*Fruitbody* closely appressed to the woody substrate, up to 16 in/40 cm in diameter, and lacking a free margin, initially forming small, round patches about 1/16 in/0.1 cm thick; surface smooth to warty, cream to olive-brown or dark brown, with a whitish, finely fringed edge. *Flesh* membranous, fibrous. *Spore deposit* olive-brown. *Habitat* on dead wood, mainly of conifers; often found in damp buildings. *Edibility* inedible.

*CONIOPHORA PUTEANA*

**CELLAR FUNGUS or WET ROT FUNGUS**

**GROUP** Bracket-fungi

**FAMILY** Dry and Wet Rot (Coniophoraceae)

**SEASON** **EDIBILITY**

**SIMILAR SPECIES**

*C. arida* has thinner fruitbodies; inedible.

## PENIOPHORA QUERCINA

One of several closely related parchment-fungi, of which this species is characterized by the reddish to ochre surface, and its habitat on beech or oak trees.

*Fruitbody* appressed to the bark substrate, spreading from initially small round patches; about 1⁄32 in/0.5 mm thick, dry and tough, reddish drying to ochre yellow, turning grey when old, smooth then eventually cracking, with a paler margin which finally rolls back to reveal a blackish brown upper surface. *Flesh* thin, whitish, tough. *Spore deposit* pale pink. *Habitat* on dead branches of beech and oak, sometimes on birch or ash. *Edibility* inedible.

*PENIOPHORA QUERCINA*

**OAK PARCHMENT or HAIRY PARCHMENT**

GROUP Bracket-fungi

FAMILY Corticium (Corticiaceae)

SEASON EDIBILITY

SIMILAR SPECIES

*P. rufomarginata* forms thicker fruitbodies and grows on linden trees; inedible.

## HYPHODERMA SAMBUCI

A common and conspicuous parchment-fungus, resembling patches of white-wash on dead branches.

*Fruitbody* closely applied to the bark, forming pure white to creamy white patches, 0.1–0.3 mm thick, finally spreading to over 4 in/10 cm in diameter, smooth or slightly warty, cracking in dry weather, with a well-defined but never free margin. *Flesh* thin, white, of chalky consistency. *Spore deposit* white. *Habitat* although often on elder wood, it can occur on many deciduous trees, often found in hedgerows. *Edibility* inedible.

*HYPHODERMA SAMBUCI*

**WHITE CRUST**

GROUP Bracket-fungi

FAMILY Corticium (Corticiaceae)

SEASON EDIBILITY

SIMILAR SPECIES

The striking white fruitbodies are distinctive.

## PULCHERRICIUM CAERULEUM

An easy to recognize parchment-fungus, owing to the striking, bright blue colours.

*Fruitbody* 1¼–3 in/3–8 cm in diameter, crust-like, about 1⁄32 in/0.5 mm thick, more or less circular but neighbouring fruitbodies may become joined; surface bright indigo blue or darker at the centre, with the margin lifted in older specimens. *Flesh* membranous, soft and waxy. *Spore deposit* whitish. *Habitat* on underside of rotting branches of ash, hazel and ivy. *Edibility* inedible.

*PULCHERRICIUM CAERULEUM*

**VELVET BLUE SPREAD**

GROUP Bracket-fungi

FAMILY Corticium (Corticiaceae)

SEASON EDIBILITY

SIMILAR SPECIES

## CHONDROSTEREUM PURPUREUM

A distinctive stereum-species, with a smooth, purplish lower surface; often attacks apple and plum trees.

*Fruitbody* closely applied to the woody substrate, with the edges curving to form small, thin caps, ¾–1½ in/2–4 cm in diameter. *Upper surface* greyish white to pale brownish, hairy, poorly zoned, with a paler, thin, wavy margin. *Lower surface* smooth, pinkish violet to purplish brown, rather waxy. *Flesh* up to 1⁄16 in/0.2 cm thick, leathery, whitish. *Spore deposit* white. *Habitat* on dead or dying wood of deciduous trees, especially apple or plum, sometimes conifers, or on wood piles. *Edibility* inedible.

*CHONDROSTEREUM PURPUREUM*

**SILVER LEAF FUNGUS**

**GROUP** Bracket-fungi

**FAMILY** Crust Fungus (Stereaceae)

**SEASON** **EDIBILITY**

**SIMILAR SPECIES**
The purplish lower surface distinguishes this species from other "stereums".

## STEREUM HIRSUTUM

The most common *Stereum* species, forming many yellowish brackets growing in tiers on dead branches.

*Cap* ¾–2¼ in/2–6 cm in diameter, bracket-like, with the broad basal attachment region spreading over the substrate. *Upper surface* yellowish orange to greyish white, hairy but less so when old, with conspicuous concentric zoning, and a thin, wavy margin. *Lower surface* smooth, yellowish to brownish orange, not bruising reddish. *Flesh* 1⁄16 in/0.1–0.2 cm thick, yellowish, leathery-tough. *Spore deposit* white. *Habitat* on dead wood of deciduous trees, especially on the wood of oak, beech and alder. *Edibility* inedible

*STEREUM HIRSUTUM*

**YELLOW STEREUM**

**GROUP** Bracket-fungi

**FAMILY** Crust Fungus (Stereaceae)

**SEASON** **EDIBILITY**

**SIMILAR SPECIES**
*S. subtomentosum* has a bright yellow margin, and grows on willow and alder; inedible. *S. insignitum* has brown zoning and grows on beech; inedible. *Coriolus versicolor* can look very similar but has white pores on the lower surface; inedible.

## STEREUM SANGUINOLENTUM

One of the *Stereum* species in which the lower surface is seen to bleed reddish when bruised, and which is always found growing on coniferous trees.

*Fruitbody* crust-like, 4–8 in/10–20 cm across, forming many, small, overlapping brackets, ¾–1¼ in/2–3 cm in diameter. *Upper surface* with reddish brown, concentric zones, finely hairy, and with a whitish margin. *Lower surface* smooth, yellowish grey to ochre-brown, sometimes with a violet tint, bruising blood-red. *Flesh* thin, whitish, leathery, drying hard. *Spore deposit* white. *Habitat* on dead wood of fir, pine, spruce and larch. *Edibility* inedible.

*STEREUM SANGUINOLENTUM*

**BLEEDING CONIFER PARCHMENT**

**GROUP** Bracket-fungi

**FAMILY** Crust Fungus (Stereaceae)

**SEASON** **EDIBILITY**

**SIMILAR SPECIES**
Other bleeding species include *S. gausapatum* which forms rusty brown brackets growing on oak, and *S. rugosum* which is more crust-like, stratified and grows on various deciduous trees; both inedible.

### *HYMENOCHAETE RUBIGINOSA*

**A** thin, brown-fleshed crust-fungus, growing on oak or chestnut, and with a smooth, reddish brown lower surface.

*Cap* ⅜–1½ in/1–4 cm in diameter, bracket-like, having a dark reddish brown to blackish upper surface, with narrow, concentric ridges, and a thin, paler, wavy margin. *Lower surface* smooth to warty, reddish brown, finally becoming greyish brown. *Flesh* brown, thin, leathery and tough. *Habitat* on dead wood of oak or chestnut. *Edibility* inedible.

*HYMENOCHAETE RUBIGINOSA*

**RED-BROWN CRUST**

**GROUP** Bracket-fungi

**FAMILY** Brown Polypore (Hymenochaetaceae)

**SEASON** **EDIBILITY**

**SIMILAR SPECIES**

*H. tabacina* is yellowish brown, not reddish brown, and grows on hazel and willow; inedible.

### *LYCOPERDON PYRIFORME*

**A** common species distinguished by the clustered, pear-shaped fruitbodies which grow on wood and have white, cord-like mycelium at the base.

*Fruitbody* 1½–3 in/4–8 cm high, ¾–1½ in/2–4 cm across the upper part, pear-shaped, usually with a raised centre; whitish at first becoming pale brown, with white, branching cord-like mycelium at the base. *Surface* scurfy at first, comprising tiny warts and granules which are soon lost, leaving a smooth inner wall. *Inner wall* thin, papery, opening by a small, irregular pore at the top. *Fertile tissue* at first white, soon greenish yellow and finally olive-brown as the spores develop, with a large, conspicuous central sterile column (the columella). *Sterile base* spongy, composed of small cells. *Habitat* in clusters on old stumps and logs or attached to buried wood. *Edibility* edible while young and still white inside.

*LYCOPERDON PYRIFORME*

**STUMP PUFFBALL or PEAR-SHAPED PUFFBALL**

**GROUP** Puffballs

**FAMILY** Puffball (Lycoperdaceae)

**SEASON** **EDIBILITY**

**SIMILAR SPECIES**

*L. lividum* is less pear-shaped, and does not occur on wood; edible.

### *CYATHUS STRIATUS*

**A** distinctive species recognized by the brown, externally shaggy fruitbodies which are distinctly grooved on the inner surface.

*Fruitbody* ¼–⅝ in/0.7–1.5 cm high, ¼–5⁄16 in/0.6–0.9 cm across, at first closed by a whitish membrane which ruptures and is soon lost at maturity; beaker-shaped, not or only slightly flared at the margin, narrowed downwards; containing several egg-shaped fertile structures (peridioles). *Outer surface* reddish brown, densely covered with coarse, shaggy tufts of hairs. *Inner surface* pale, greyish or grey-brown, smooth, distinctly longitudinally grooved or furrowed. *Peridioles* ovoid, whitish or pale greyish, smooth, attached by a slender, whitish thread. *Habitat* in clusters, attached to twigs, leaves or other debris, often in gardens or on disturbed ground. *Edibility* inedible.

*CYATHUS STRIATUS*

**GROOVED BIRD'S NEST FUNGUS or SPLASH CUPS**

**GROUP** Puffballs

**FAMILY** Bird's Nest Fungi (Nidulariaceae)

**SEASON** **EDIBILITY**

**SIMILAR SPECIES**

*C. olla* has trumpet-shaped fruitbodies which are paler, felty on the surface and lack a striated inner surface; inedible.

## CRUCIBULUM LAEVE

Recognized by the short cylindric or cup-shaped fruitbody with a brown outer surface, and gregarious growth on litter.

*Fruitbody* $\frac{3}{16}$–$\frac{3}{8}$ in/0.5–1 cm high, $\frac{3}{16}$–$\frac{5}{16}$ in/0.4–0.9 cm across, cup- or beaker-shaped, slightly flared at the margin; at first globose and closed by a pale yellowish membrane which is soon ruptured and lost; containing several, tiny, egg-like structures. *Outer surface* downy at first, becoming smooth, tawny-brown. *Inner surface* whitish or pale grey-brown, shiny, smooth. *Peridioles* $\frac{1}{16}$ in/0.1–0.2 cm in diameter, lens-shaped, whitish or pale yellowish brown, each attached by a slender whitish thread. *Habitat* gregarious on dead twigs, stems and other debris. *Edibility* inedible.

*CRUCIBULUM LAEVE*

**COMMON BIRD'S NEST or WHITE-EGG BIRD'S NEST**

**GROUP** Puffballs

**FAMILY** Bird's Nest Fungi (Nidulariaceae)

**SEASON** **EDIBILITY**

**SIMILAR SPECIES**

*C. olla* is much more flared at the margin, paler in colour and more tapered at the base; inedible.

## CALOCERA VISCOSA

A common species recognized by the branched, orange-yellow, tough gelatinous fruitbodies which grow on conifer stumps.

*Fruitbody* $\frac{3}{4}$–3 in/2–8 cm high, bright orange-yellow, deeper orange when dry; erect, repeatedly branched, with a rooting base; surface smooth, rather slimy. *Flesh* tough, gelatinous, yellow. *Spore deposit* yellow. *Habitat* on old stumps of coniferous trees, especially pines. *Edibility* inedible.

*CALOCERA VISCOSA*

**JELLY ANTLER FUNGUS or YELLOW TUNING FORK**

**GROUP** Jelly Fungi

**FAMILY** Jelly Antler Fungus (Dacrymycetaceae)

**SEASON** **EDIBILITY**

**SIMILAR SPECIES**

Other species of *Calocera* are smaller, and have simple or sparsely branched fruitbodies. Species of *Clavaria* and allied genera differ in having brittle, non-gelatinous fruitbodies. All inedible.

## EXIDIA GLANDULOSA

A common and distinctive species recognized by the blackish, disc-shaped or irregular, gelatinous fruitbodies with warted fertile surface.

*Fruitbody* $\frac{3}{8}$–2 in/1–5 cm across, sometimes confluent and larger, pendulous, gelatinous, blackish throughout; variable in shape, disc-shaped or top-shaped at first, soon irregular. *Fertile surface* smooth or irregularly folded, bearing numerous small, wart-like projections. *Sterile surface* adjacent to the substratum, minutely downy. *Spore deposit* white. *Habitat* on dead logs and stumps of deciduous trees, especially oak. *Edibility* inedible.

*EXIDIA GLANDULOSA*

**BLACK WITCHES' BUTTER or BLACK JELLY ROLL**

**GROUP** Jelly Fungi

**FAMILY** Yellow Brain Fungus (Tremellaceae)

**SEASON** **EDIBILITY**

**SIMILAR SPECIES**

*Bulgaria inquinans* is black and gelatinous, but has a blackish spore print; inedible.

## *TREMELLA MESENTERICA*

**A** distinctive species recognized by the orange-yellow colour, gelatinous flesh and irregular, brain-like shape.

*Fruitbody* ⅜–3 in/1–8 cm across, bright orange-yellow, irregular in shape, folded and lobed, softly gelatinous throughout, drying dark orange and horny. *Fertile surface* continuous. *Spore deposit* white. *Habitat* on dead branches of deciduous trees; common. *Edibility* inedible.

*TREMELLA MESENTERICA*

**YELLOW BRAIN FUNGUS or WITCHES' BUTTER**

GROUP Jelly Fungi

FAMILY Yellow Brain Fungus (Tremellaceae)

SEASON EDIBILITY

SIMILAR SPECIES

Unlikely to be confused with other species.

## *AURICULARIA AURICULA-JUDAE*

**A** distinctive species recognized by the gelatinous, ear-shaped fruitbodies growing on dead wood.

*Fruitbody* 1¼–4 in/3–10 cm across, irregularly cup-shaped or ear-shaped, reddish brown, gelatinous, drying hard; laterally attached without a stalk, the outer surface covered with short, greyish hairs. *Fertile surface* greyish brown, usually wrinkled or veined, otherwise smooth. *Flesh* thin, slightly translucent; no distinctive smell. *Spore deposit* white. *Habitat* on dead branches, especially of elder and elm, usually gregarious; common. *Edibility* edible.

*AURICULARIA AURICULA-JUDAE*

**JEW'S EAR or TREE EAR**

GROUP Jelly Fungi

FAMILY Jew's Ear (Auriculariaceae)

SEASON EDIBILITY

SIMILAR SPECIES

When mature, unlikely to be confused with other species.

## *AURICULARIA MESENTERICA*

**A** common species distinguished by the gelatinous, bracket-like fruitbodies which have a densely hairy, zoned upper surface.

*Fruitbody* 1½–4 in/4–10 cm across, at first cup-shaped, soon widely spreading over the substratum, with reflexed, undulating, bracket-like portions. *Upper surface* pale greyish or whitish, shaggy with dense, whitish hairs, concentrically zoned. *Fertile surface* reddish purple, often with a whitish bloom, smooth, veined. *Flesh* firm, gelatinous, horny when dry. *Spore deposit* white. *Habitat* on stumps and logs of deciduous trees, growing especially on elm; common. *Edibility* inedible.

*AURICULARIA MESENTERICA*

**TRIPE FUNGUS**

GROUP Jelly Fungi

FAMILY Jew's Ear (Auriculariaceae)

SEASON EDIBILITY

SIMILAR SPECIES

*Chondrostereum purpureum* has similar, shaggy brackets and purplish fertile surface, but is not gelatinous; inedible.

## SCUTELLINIA SCUTELLATA

Recognized by the gregarious, bright orange-red, disc-shaped fruitbodies which bear long, dark brown hairs at the margin.

*Fruitbody* 1/8–3/8 in/0.3–1 cm across, saucer-shaped, slightly concave, bearing stiff, dark brown tapered hairs up to 1/16 in/0.1 cm long on the outer surface, especially at the margin. *Fertile surface* bright orange-red, smooth. *Outer surface* pale brown. *Flesh* pale orange. *Habitat* gregarious on rotten wood or damp soil; common. *Edibility* inedible.

*SCUTELLINIA SCUTELLATA*

**EYELASH FUNGUS or EYELASH CUP**

**GROUP** Cup Fungi

**FAMILY** Eyelash Cup Fungi Pyronemataceae

**SEASON** **EDIBILITY**

**SIMILAR SPECIES**

Other species of *Scutellinia* may be similar; many have shorter marginal hairs, but microscopic examination is required for certain identification. All species inedible.

## BULGARIA INQUINANS

A common species, which is recognized by the blackish, thick-fleshed, rubbery, clustered fruitbodies.

*Fruitbody* 3/8–1½ in/1–4 cm across, gelatinous, with a rubbery texture, at first subglobose, inrolled at the margin, later expanding, top-shaped. *Fertile surface* black, smooth, flattened or slightly concave. *Outer surface* brownish, scurfy. *Flesh* thick, gelatinous, blackish. *Spore deposit* blackish. *Habitat* clustered, on dead trunks of deciduous trees, especially oak. *Edibility* inedible.

*BULGARIA INQUINANS*

**BLACK BULGAR or BLACK JELLY DROPS**

**GROUP** Cup Fungi

**FAMILY** Jelly Discs (Leotiaceae)

**SEASON** **EDIBILITY**

**SIMILAR SPECIES**

*Exidia glandulosa* may be similar in form, but has small warts on the fertile surface and a white spore print; inedible.

## DALDINIA CONCENTRICA

A distinctive species having blackish, hemispherical fruitbodies of which the flesh has a distinctive concentric zonation.

*Fruitbody* 3/4–3 in/2–8 cm across, subglobose or hemispherical, at first reddish brown, soon becoming black, shiny, smooth. *Flesh* rather fibrous, dark purplish brown or grey-brown, with conspicuous darker concentric zones. *Spore deposit* black. *Habitat* gregarious, on dead or dying deciduous trunks, especially ash and beech, also on charred birch; common. *Edibility* inedible.

*DALDINIA CONCENTRICA*

**CRAMP BALLS or KING ALFRED'S CAKES or CARBON BALLS**

**GROUP** Flask Fungi

**FAMILY** Candle Snuff (Xylariaceae)

**SEASON** **EDIBILITY**

**SIMILAR SPECIES**

*D. vernicosa* has smaller fruitbodies which occur on burnt gorse; inedible.

## *HYPOXYLON MULTIFORME*

**A** common species recognized by the red-brown then blackish, distinctly papillate fruitbodies which grow on birch and alder.

*Fruitbody* ½–1½ in/1–3 cm across, cushion-shaped or irregular in form, often developing an extensive crust, at first reddish-brown, soon becoming black, the surface distinctly papillate. *Flesh* ⅛–¼ in/0.3–0.7 cm thick, black, brittle. *Spore deposit* black. *Habitat* on dead branches of birch and alder; common. *Edibility* inedible.

*HYPOXYLON MULTIFORME*

### BIRCH CUSHION

**GROUP** Flask Fungi

**FAMILY** Candle Snuff (Xylariaceae)

**SEASON** **EDIBILITY**

**SIMILAR SPECIES**

Other species of *Hypoxylon* with initially reddish fruitbodies are less papillate, but some may require microscopic identification.

## *XYLARIA POLYMORPHA*

**A** large, black, club-shaped species which grows in clusters on dead stumps.

*Fruitbody* 1¼–4 in/3–10 cm high, ⅜–1¼ in/1–3 cm across, black, variable in form, irregularly club-shaped, sometimes lobed, with a short, cylindrical stalk; surface of fertile part minutely papillate, granular. *Flesh* white, with a thin black crust in which the fertile, flask-shaped structures (perithecia) are embedded. *Spore deposit* black. *Habitat* in clusters on dead stumps, especially of beech. *Edibility* inedible.

*XYLARIA POLYMORPHA*

### DEAD MAN'S FINGERS

**GROUP** Flask Fungi

**FAMILY** Candle Snuff (Xylariaceae)

**SEASON** **EDIBILITY**

**SIMILAR SPECIES**

*X. longipes* is more slender, less clustered and occurs on roots and branches of sycamore and other species of *Acer*; inedible.

## *XYLARIA HYPOXYLON*

**A** common species recognized by the slender, erect, usually branched fruitbodies which are either black or have a white, powdery coating to the upper part.

*Fruitbody* ¾–3 in/2–8 cm high, ⅛–¼ in/0.3–0.6 cm across, erect, cylindrical to flattened, usually forked and antler-shaped, pointed, stalked; upper fertile portion at first white, powdery, later darkening, becoming black and papillate. Unbranched, black, cylindrical fruitbodies are also commonly present. *Stalk* black, hairy. *Flesh* white, with a thin, black crust. *Spore deposit* black. *Habitat* on dead wood of various kinds; common. *Edibility* inedible.

*XYLARIA HYPOXYLON*

### CANDLE SNUFF FUNGUS or CARBON ANTLERS

**GROUP** Flask Fungi

**FAMILY** Candle Snuff (Xylariaceae)

**SEASON** **EDIBILITY**

**SIMILAR SPECIES**

*X. carpophila* is more slender and occurs on rotten beech mast in litter; inedible.

CHAPTER FIVE

# IN WET SITUATIONS, SUCH AS BOGS AND MARSHLANDS

## *TEPHROCYBE PALUSTRE*

**A** slender toadstool, growing in small groups in damp situations, and having a greyish brown cap and grey gills.

*Cap* ⅜–1¼ in/1–3 cm in diameter, conical but soon becoming convex, dark greyish brown, drying paler, striated at the margin. *Gills* adnexed, pale grey, widely spaced. *Stem* 1½–4¼ × ⅛–3/16 in/4–11 × 0.3–0.5 cm, tall and slender, similarly coloured to the cap, smooth. *Flesh* thin, watery grey, with a strong odour of cucumber. *Spore deposit* white. *Habitat* among *Sphagnum* moss, in wet localities. *Edibility* inedible.

*TEPHROCYBE PALUSTRE*

**SWAMP GREY GILL**

**GROUP** Mushrooms and Toadstools

**FAMILY** Tricholoma (Tricholomataceae)

**SEASON** **EDIBILITY**

**SIMILAR SPECIES**

Unlikely to be confused with other species.

## *HYPHOLOMA ELONGATUM*

**C**ommonly found in large groups among *Sphagnum* moss, and recognized by the yellowish cap and stem.

*Cap* ⅜–1¼ in/1–3 cm in diameter, bell-shaped becoming flattened, honey-yellow, sometimes with a greenish tint, smooth, with a straight, striated margin. *Gills* adnate, greyish lilac, finally becoming dark brown, crowded. *Stem* 2–4 × ⅛–3/16 in/5–10 × 0.3–0.4 cm, tall and slender, hollow, yellowish brown, more whitish towards the apex, smooth. *Flesh* thin, white, brittle. *Spore deposit* sooty brown. *Habitat* among *Sphagnum* moss, growing in marshland. *Edibility* inedible.

*HYPHOLOMA ELONGATUM*

**SWAMP SULPHUR CAP**

**GROUP** Mushrooms and Toadstools

**FAMILY** Stropharia (Strophariaceae)

**SEASON** **EDIBILITY**

**SIMILAR SPECIES**

*Pholiota myosotis* has a brown cap and a brown spore deposit; inedible.

## *HYPHOLOMA UDUM*

**A** fairly common species found in *Sphagnum* moss, distinguished by a non-sticky cap and greyish brown gills.

*Cap* ⅜–¾ in/1–2 cm in diameter, conical, expanding but usually retaining a raised centre, yellowish brown, not sticky, smooth, with a striated margin. *Gills* adnate, greyish brown, finally dark brown, moderately crowded. *Stem* 1½–3½ × ⅛–3/16 in/4–9 × 0.3–0.5 cm, tall and slender, cylindrical, similarly coloured to the cap. *Flesh* thin, whitish, brittle. *Spore deposit* greyish brown. *Habitat* among peat of *Sphagnum* moss. *Edibility* inedible.

*HYPHOLOMA UDUM*

**PEAT SULPHUR CAP**

**GROUP** Mushrooms and Toadstools

**FAMILY** Stropharia (Strophariaceae)

**SEASON** **EDIBILITY**

**SIMILAR SPECIES**

*Pholiota myosotis* has a sticky cap and a yellowish brown spore deposit; inedible.

## DERMOCYBE ULIGINOSA

**A** bright orange-brown toadstool which has rusty brown gills and grows on muddy ground under willow and alder trees.

*Cap* 1¼–3 in/3–8 cm in diameter, convex with a raised centre, eventually becoming depressed, bright tawny orange, smooth and dry. *Gills* adnexed to adnate, lemon-yellow, finally becoming rusty brown, crowded. *Stem* 1¼–4 × 3/16–3/8 in/3–10 × 0.5–1 cm, stocky, similarly coloured to the cap, more yellowish towards the apex, striated, and with a yellow, cobweb-like veil in the young stages. *Flesh* lemon-yellow, with an odour of radish. *Spore deposit* rusty brown. *Habitat* in wet, boggy areas, especially under alder or willow, and often among *Sphagnum* moss. *Edibility* inedible, and easily confused with poisonous *Cortinarius* species.

*DERMOCYBE ULIGINOSA*

### SWAMP DERMOCYBE

**GROUP** Mushrooms and Toadstools

**FAMILY** Cortinarius (Cortinariaceae)

**SEASON** **EDIBILITY**

**SIMILAR SPECIES**

*Cortinarius palustris* has a greenish brown cap and stem; poisonous.

## GALERINA TIBIICYSTIS

**An** early species recognized by habitat, bright orange-brown colours, powdery stem and lack of smell.

*Cap* 5/8–1¼ in/1.5–3 cm across, convex or bell-shaped at first, soon expanding, usually with a raised centre, reddish- or orange-brown, drying paler, margin weakly striated. *Gills* concolorous with the cap, adnexed. *Stem* 2¼–3½ × 1/8–3/16 in/6–9 × 0.3–0.4 cm, slender, cylindrical, similarly coloured to the cap or paler, sometimes whitish at the base, minutely powdery over the entire surface. *Flesh* pale orange-brown, without a distinctive smell. *Spore deposit* brown. *Habitat* in swampy areas with bog-mosses, *Sphagnum* species. *Edibility* inedible.

*GALERINA TIBIICYSTIS*

### BOG-MOSS HELMET CAP or SPHAGNUM-BOG GALERINA

**GROUP** Mushrooms and Toadstools

**FAMILY** Cortinarius (Cortinariaceae)

**SEASON** **EDIBILITY**

**SIMILAR SPECIES**

*G. sphagnorum*, in similar habitats, has a distinctive mealy taste and smell; inedible.

## NAUCORIA ESCHAROIDES

**P**ale colours and occurrence in damp places under alders distinguish this common species.

*Cap* 5/16–1 in/0.8–2.5 cm across, pale buff or milky-coffee colour, convex or flattened, slightly scurfy, not striate at the margin when moist. *Gills* adnate, narrow, crowded, ochraceous or pale tan. *Stem* 5/8–1½ × 1/16–1/8 in/1.5–4 × 0.1–0.3 cm, similarly coloured to the cap or darker below, slender, cylindrical, often slightly curved, with a smooth surface. *Flesh* thin, pale brown or yellowish, becoming hollow in the stem, with a slightly sweetish smell and radishy taste. *Spore deposit* brown. *Habitat* gregarious, on damp ground under alders; frequent. *Edibility* inedible.

*NAUCORIA ESCHAROIDES*

### CAMPANULATE NAUCORIA

**GROUP** Mushrooms and Toadstools

**FAMILY** Cortinarius (Cortinariaceae)

**SEASON** **EDIBILITY**

**SIMILAR SPECIES**

Several other species of *Naucoria* occur under alder. These are less common and usually darker in colour, but require microscopic examination for positive identification.

## LEOTIA LUBRICA

**A** distinctive species recognized by the gelatinous fruitbody with yellowish stalk and olive-yellow head.

*Fruitbody* $\frac{3}{4}$–$2\frac{1}{4}$ in/2–6 cm high, capitate, gelatinous. *Fertile head* $\frac{3}{16}$–$\frac{3}{8}$ in/0.4–1 cm across, convex, irregularly lobed at the margin, yellowish green or olivaceous, smooth, rather slimy. *Stalk* $\frac{1}{8}$–$\frac{5}{16}$ in/0.3–0.8 cm wide, cylindrical, tapered at the base, ochraceous, covered in tiny, greenish granules. *Flesh* yellowish, gelatinous, becoming hollow in the stem, without a distinctive smell. *Spore deposit* white. *Habitat* on damp ground in deciduous woodland, gregarious or clustered. *Edibility* inedible.

*LEOTIA LUBRICA*

**JELLY BABIES or OCHRE JELLY CLUB**

**GROUP** Cup Fungi

**FAMILY** Jelly Discs (Leotiaceae)

**SEASON** **EDIBILITY**

**SIMILAR SPECIES**

*L. atrovirens*, in North America, differs in being wholly green, and *L. viscosa*, also in North America, has a dark green head; both inedible.

## MITRULA PALUDOSA

**R**ecognized by the orange-yellow head and white stem, and occurrence in wet places.

*Fruitbody* $\frac{5}{8}$–2 in/1.5–5 cm high, erect, slender. *Fertile head* ovoid or cylindrical, sometimes furrowed at the base, bright yellowish orange, smooth, clearly delimited from the stem. *Stem* $\frac{1}{16}$–$\frac{1}{8}$ in/0.1–0.3 cm wide, white, smooth, often covered in algae and debris at the base. *Flesh* whitish, soft. *Habitat* gregarious in damp ditches and wet areas, among leaf-litter or in *Sphagnum* moss. *Edibility* inedible.

*MITRULA PALUDOSA*

**BOG BEACON**

**GROUP** Cup Fungi

**FAMILY** Earth Tongue (Geoglossaceae)

**SEASON** **EDIBILITY**

**SIMILAR SPECIES**

Unlikely to be confused with other species.

CHAPTER SIX

# ON BURNT GROUND OR ON BURNT WOOD

## *PHOLIOTA HIGHLANDENSIS*

A very common toadstool on burnt ground and charred wood, distinguished by the slimy cap and reddish brown gills.

*Cap* 1¼–2 in/3–5 cm in diameter, convex becoming flattened with a wavy margin, chestnut-brown becoming paler on drying, slightly sticky, smooth. *Gills* adnate, clay-brown, crowded. *Stem* ¾–2 × ⅛–³⁄₁₆ in/2–5 × 0.3–0.5 cm, short, yellowish brown, paler at the apex, with small fibre-scales. *Flesh* thin, pale, firm. *Spore deposit* cinnamon-brown. *Habitat* on burnt ground or burnt stumps. *Edibility* inedible.

*PHOLIOTA HIGHLANDENSIS*

**CHARCOAL PHOLIOTA**

**GROUP** Mushrooms and Toadstools

**FAMILY** Stropharia (Strophariaceae)

**SEASON** **EDIBILITY**

**SIMILAR SPECIES**

*P. spumosa* has a slimy, yellowish cap, and grows in conifer woods; inedible.

## *TEPHROCYBE ATRATA*

A small toadstool, often forming small tufts on burnt ground, recognized by the blackish cap and grey gills.

*Cap* ¾–2 in/2–5 cm in diameter, convex finally becoming depressed at the centre, black when moist but drying out to become pale brown, smooth. *Gills* adnate, pale grey, spaced. *Stem* ¾–2 × ⅛–³⁄₁₆ in/2–5 × 0.3–0.5 cm, slender, sooty brown, smooth. *Flesh* thin, brown, with an odour of damp flour or rancid oil. *Spore deposit* white. *Habitat* on burnt soil and charcoal, in small groups. *Edibility* worthless.

*TEPHROCYBE ATRATA*

**CHARCOAL COLLYBIA**

**GROUP** Mushrooms and Toadstools

**FAMILY** Tricholoma (Tricholomataceae)

**SEASON** **EDIBILITY**

**SIMILAR SPECIES**

*T. ambusta* is more slender, with a striated cap margin; *Myxomphalia maura* has decurrent gills and grows on burnt ground in conifer woods. Both species are inedible.

## *RHIZINA UNDULATA*

Forms distinctive, dark brown, convex fruitbodies having a pale margin and firm flesh, and attached to the substrate by whitish root-like structures.

*Fruitbody* 1½–4¾ in/4–12 cm across, convex and cushion-like, often lobed and irregular in outline. *Upper surface* undulating, dark chestnut-brown, with a pale, cream margin, smooth. *Underside* cream or pale ochraceous, attached to the substrate by numerous branched, whitish, root-like structures. *Flesh* firm, tough, thick, reddish brown, without a distinctive smell. *Habitat* usually on burnt ground in coniferous woods; causes a disease of conifers known as group dying. *Edibility* inedible.

*RHIZINA UNDULATA*

**PINE FIRE FUNGUS or CRUSTLIKE CUP**

**GROUP** Cup Fungi

**FAMILY** Saddle-cup (Helvellaceae)

**SEASON** **EDIBILITY**

**SIMILAR SPECIES**

Unlikely to be confused with other species.

## *PEZIZA PROTEANA* VAR. *SPARASSOIDES*

The large, convoluted, cauliflower-like, whitish and brittle-fleshed fruitbodies are distinctive.

*Fruitbody* 1¼–4¾ in/3–12 cm in diameter, 3–8 in/8–20 cm high, much folded and convoluted and rather cauliflower-like in appearance, cream or pale ochraceous, sometimes with a pinkish tinge. *Fertile surface* smooth. *Flesh* whitish, brittle, without a distinctive smell. *Habitat* on burnt ground, uncommon. *Edibility* edible after cooking, but poor.

*PEZIZA PROTEANA* VAR. *SPARASSOIDES*

**CONVOLUTED BONFIRE CUP**

**GROUP** Cup Fungi

**FAMILY** Large Cup Fungi (Pezizaceae)

**SEASON** **EDIBILITY**

**SIMILAR SPECIES**

Species of *Sparassis* may be similar in general appearance, but have tougher flesh and differ in habitat.

## *PEZIZA VIOLACEA*

Recognized by the cup or disc-shaped, violaceous fruitbodies which occur growing on burnt ground.

*Fruitbody* ⅜–1½ in/1–4 cm across, cup-shaped, expanding and becoming saucer-shaped, irregular in outline when clustered, lacking a stalk. *Fertile surface* violaceous, sometimes more brownish when old, smooth. *Outer surface* paler, slightly scurfy. *Flesh* thin, violaceous, without a distinctive smell. *Habitat* on burnt ground, fairly common. *Edibility* inedible.

*PEZIZA VIOLACEA*

**VIOLET BONFIRE CUP**

**GROUP** Cup Fungi

**FAMILY** Large Cup Fungi (Pezizaceae)

**SEASON** **EDIBILITY**

**SIMILAR SPECIES**

*P. pseudoviolacea* is similar, and can be distinguished with certainty only by microscopic examination; inedible.

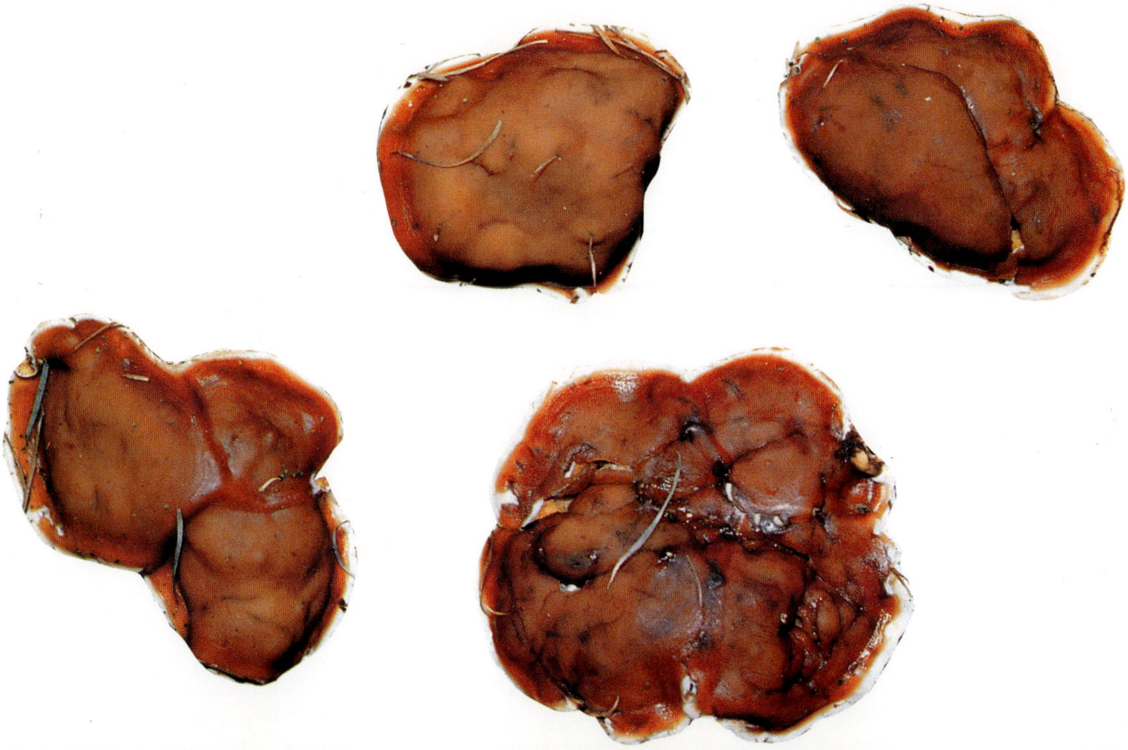

CHAPTER SEVEN

# ON DUNG OR ENRICHED SOIL

## *VOLVARIELLA SPECIOSA*

**A** tall, solitary, pale mushroom with a sticky cap, pink gills, and a sac-like volva.

*Cap* 2¾–5½ in/7–14 cm in diameter, conical to convex, finally becoming flattened, whitish with a pale greyish brown centre, smooth, slimy when moist. *Gills* free, at first white gradually becoming pink, thin, broad, and very crowded. *Stem* 3½–7 × ⅜–⅝ in/9–18 × 1–1.5 cm, tall and cylindrical, fragile, hollow, whitish, with fine, brownish streaks; arising from a sac-like, white volva, which soon collapses. *Flesh* thin, soft and soon decaying. *Spore deposit* salmon-pink. *Habitat* growing on compost heaps, and richly manured soil. *Edibility* said to be edible but great care must be taken to avoid confusion with the poisonous, white-gilled *Amanita* species.

*VOLVARIELLA SPECIOSA*

**ROSE-GILLED GRISETTE or SMOOTH VOLVARIELLA**

**GROUP** Mushrooms and Toadstools

**FAMILY** Pluteus (Pluteaceae)

**SEASON** **EDIBILITY**

**SIMILAR SPECIES**

Pink-gilled *Pluteus* species (edible, but worthless) lack a volva, while the poisonous *Amanita* species have white gills.

## *PANAEOLUS SEMIOVATUS*

**A** distinctive species with mottled gills, a pale-coloured cap and a ring on the stem.

*Cap* ¾–1½ in/2–4 cm in diameter, bell-shaped and remaining so, whitish to pale brown, smooth and sticky, finally shining when dry. *Gills* adnate, crowded, grey becoming black, mottled, with a white edge. *Stem* 2–4¾ × ⅛–3/16 in/5–12 × 0.3–0.5 cm, tall and slender, whitish above, yellowish brown below, bearing a conspicuous, white, membranous ring attached towards the middle. *Flesh* thin, white, brittle, with an odour of rotting straw. *Spore deposit* black. *Habitat* frequently found growing on dung, especially of cow and horse. *Edibility* inedible, possibly poisonous.

*PANAEOLUS SEMIOVATUS*

**DUNG MOTTLE-GILL**

**GROUP** Mushrooms and Toadstools

**FAMILY** Ink-cap (Coprinaceae)

**SEASON** **EDIBILITY**

**SIMILAR SPECIES**

*P. phalaenarum* is more slender and lacks a ring; inedible.

## *PANAEOLUS SPHINCTRINUS*

**O**ne of several, slender mottle-gills, recognized by a dark grey stem, and veil scales on the cap margin.

*Cap* ¾–1¼ in/2–3 cm in diameter, bell-shaped and remaining so, dark grey, smooth, often with a fringe of tiny, white scales of the veil attached to the margin. *Gills* adnate, blackish brown, mottled, crowded. *Stem* 2¾–4¾ × 1/16–⅛ in/7–12 × 0.2–0.3 cm, tall and slender, dark greyish to blackish, finely powdery. *Flesh* thin, grey, firm. *Habitat* common, especially growing on cowpats in fields. *Edibility* inedible, possibly poisonous.

*PANAEOLUS SPHINCTRINUS*

**GREY MOTTLE-GILL**

**GROUP** Mushrooms and Toadstools

**FAMILY** Ink-cap (Coprinaceae)

**SEASON** **EDIBILITY**

**SIMILAR SPECIES**

*P. campanulatus* has a reddish brown cap and stem and is usually found on horse dung; mildly poisonous.

## PSILOCYBE COPROPHILA

**A** small brown toadstool on herbivore dung, recognized by the dark brown, slimy cap and whitish, fibrillose stem.

*Cap* 3/16–¾ in/0.5–2 cm across, hemispherical, usually with a raised centre, reddish brown or tan, with a detachable, slimy layer, slightly striated at the margin. *Stem* ¾–1½ × 1/16–⅛ in/2–4 × 0.1–0.3 cm, pale brownish or whitish, rather whitish-downy near the base. *Gills* grey-brown, adnexed, crowded, broad. *Flesh* thin, pale greyish brown, darker in the lower stem, without a distinctive smell. *Spore deposit* violaceous. *Habitat* on dung of herbivores, usually gregarious. *Edibility* inedible.

*PSILOCYBE COPROPHILA*

**DUNG SMOOTH HEAD**

**GROUP** Mushrooms and Toadstools

**FAMILY** Stropharia (Strophariaceae)

**SEASON** **EDIBILITY**

**SIMILAR SPECIES**

Other small species of *Psilocybe*, such as *P. subcoprophila*, can be distinguished with certainty only by microscopic examination; inedible.

## STROPHARIA SEMIGLOBATA

**A** distinctive species recognized by the yellowish, hemispherical, slimy cap, thin ring on stem, and habitat.

*Cap* ⅜–1½ in/1–4 cm across, hemispherical, rarely expanding, pale yellow or straw coloured, slimy, smooth. *Stem* 1¼–4 × 1/16–3/16 in/3–10 × 0.2–0.4 cm, cylindrical, slightly bulbous at the base, whitish or pale yellowish, bearing a thin, fragile ring which is often incomplete or lost. *Gills* adnate, broad, purplish brown later blackish. *Flesh* pale, thin, without a distinctive smell. *Spore deposit* purplish brown. *Habitat* on dung of herbivores, especially horse and cattle, common. *Edibility* inedible.

*STROPHARIA SEMIGLOBATA*

**DUNG ROUNDHEAD or ROUND STROPHARIA**

**GROUP** Mushrooms and Toadstools

**FAMILY** Stropharia (Strophariaceae)

**SEASON** **EDIBILITY**

**SIMILAR SPECIES**

Unlikely to be confused with other species.

## PEZIZA VESICULOSA

**F**orms clusters of large, thick-fleshed, cup-shaped fruitbodies on manured soil.

*Fruitbody* 1¼–4 in/3–10 cm across, cup-shaped with a strongly inrolled, crenulate margin. *Inner surface* pale yellowish brown, usually wrinkled, or blistered in large specimens. *Outer surface* paler, coarsely scurfy. *Flesh* thick, brittle, pale fawn. *Habitat* on rich soil or manure, often in large clusters; common. *Edibility* inedible.

*PEZIZA VESICULOSA*

**EARLY CUP FUNGUS or BLADDER CUP**

**GROUP** Cup Fungi

**FAMILY** Large Cup Fungi (Pezizaceae)

**SEASON** **EDIBILITY**

**SIMILAR SPECIES**

Other species of *Peziza* are less massive and have thinner flesh, but, in order to distinguish small specimens of *P. vesiculosa*, microscopic examination may be required.

CHAPTER EIGHT

# On Other Fungi

## *NYCTALIS ASTEROPHORA*

**A** small toadstool having powdery caps and growing in clusters on very rotten fruitbodies of the Blackening Russula.

*Cap* 3/8–3/4 in/1–2 cm in diameter, flattened ball-shaped, white soon developing a powdery, fawn-coloured surface. *Gills* very reduced and virtually lacking. *Stem* 3/8–5/8 × 1/16–3/16 in/1–1.5 × 0.2–0.5 cm, short, cylindrical, white. *Flesh* thin, white, with an odour of damp meal. *Spore deposit* white. *Habitat* on rotting fruitbodies of the Blackening Russula *(Russula nigricans)*. *Edibility* worthless.

*NYCTALIS ASTEROPHORA*

**POWDER CAP**

GROUP Mushrooms and Toadstools

FAMILY Tricholoma (Tricholomataceae)

SEASON EDIBILITY

SIMILAR SPECIES

*N. parasitica* (see this page).

## *NYCTALIS PARASITICA*

**A** small, slender, pale toadstool, growing in groups on old, rotting fruitbodies of brittle-gills and milk-caps.

*Cap* 3/8–1 1/4 in/1–3 cm in diameter, bell-shaped then expanding, pale lilac-grey, smooth with a silky surface. *Gills* broadly adnate, white to brownish, often thick and distorted, well spaced. *Stem* 5/8–1 1/2 × 1/16–3/16 in/1.5–4 × 0.2–0.4 cm, cylindrical, greyish white. *Flesh* thin, white, firm. *Spore deposit* white. *Habitat* on rotting brittle-gills (*Russula* species) or milk-caps (*Lactarius* species). *Edibility* worthless.

*NYCTALIS PARASITICA*

**PICK-A-BACK TOADSTOOL**

GROUP Mushrooms and Toadstools

FAMILY Tricholoma (Tricholomataceae)

SEASON EDIBILITY

SIMILAR SPECIES

*Collybia tuberosa* (inedible) and related species grow on similar substrates but have well-developed gills. The stem of *C. tuberosa* originates from a small brown tuber (sclerotium).

## *VOLVARIELLA SURRECTA*

**A** rare species having whitish fruitbodies, with pink gills and a large volva, and occurring on rotting species of *Clitocybe*.

*Cap* 1 1/4–3 in/3–8 cm across, conical or convex, expanding, sometimes with a raised centre, whitish, sometimes greyish with age, slightly sticky at first, then silky-scaly. *Stem* 1 1/4–2 × 3/16–3/8 in/3–5 × 0.5–1 cm, cylindrical, thickened towards the base, whitish, later pale buff, slightly striate, powdery at the top, arising from a sac-like volva. *Volva* rather large, up to 5/8 in/1.5 cm across, whitish. *Gills* white at first, then salmon-pink with white edges. *Flesh* whitish, without a distinctive smell. *Habitat* clustered or tufted on old fruitbodies of *Clitocybe nebularis,* occasionally on other species of *Clitocybe* and *Tricholoma*; rare. *Edibility* inedible.

*VOLVARIELLA SURRECTA*

**PARASITIC VOLVARIA**

GROUP Mushrooms and Toadstools

FAMILY Pluteus (Pluteaceae)

SEASON EDIBILITY

SIMILAR SPECIES

Other species of *Volvariella* differ especially in habitat.

## *XEROCOMUS PARASITICUS*

The parasitic habit on the Common Earthball *(Scleroderma citrinum)* is a diagnostic character for this species.

*Cap* ¾–1½ in/2–4 cm across, convex, olive-yellow, often with a pinkish tinge, minutely velvety. *Stem* 1–1½ × ¼–⅜ in/2.5–4 × 0.6–1 cm, similarly coloured to the cap, cylindrical or tapered, arising from the underside of the host fruitbody and usually curved. *Tubes* adnate or slightly decurrent, lemon-yellow or reddish; *pores* lemon-yellow, becoming rusty with age. *Spore deposit* brownish, with olive tinge. *Habitat* parasitic on fruitbodies of the Common Earthball; uncommon. *Edibility* edible.

*XEROCOMUS PARASITICUS*

**PARASITIC BOLETE**

**GROUP** Boletes

**FAMILY** Xerocomus (Xerocomaceae)

**SEASON** **EDIBILITY**

**SIMILAR SPECIES**

Other species of *Xerocomus* may be similar in appearance, but are never parasitic on other fungi.

## *CORDYCEPS OPHIOGLOSSOIDES*

Recognized by the blackish, ovoid fertile head, yellow stem and parasitic habit on false truffles.

*Fruitbody* 2¼–4 in/6–10 cm high, club-shaped. *Fertile head* ⅝–1 × 5/16–½ in/1.5–2.5 × 0.8–1.3 cm, cylindrical or ovoid, at first yellow, smooth, becoming blackish and minutely roughened. *Stem* 1/16–⅛ in/0.2–0.3 cm wide, cylindrical, smooth, yellow. *Habitat* in woods, parasitic on the underground fruitbodies of *Elaphomyces* species. Care must be taken when collecting so that the host attachment is not lost. *Edibility* inedible.

*CORDYCEPS OPHIOGLOSSOIDES*

**SLENDER TRUFFLE CUP or GOLDENTHREAD CORDYCEPS**

**GROUP** Flask Fungi

**FAMILY** Vegetable Caterpillar (Clavicipitaceae)

**SEASON** **EDIBILITY**

**SIMILAR SPECIES**

*C. canadensis* and *C. capitatus* occur on false truffles but have larger, capitate fruitbodies; both inedible.

# A Guide To Literature On Identification


## General

**Dickinson, C & Lucas J.** *The Encyclopedia of Mushrooms.* Orbis, London, 1979.
**Hawksworth, D L, Sutton, B C & Ainsworth, G C.** *Ainsworth & Bisby's Dictionary of the Fungi,* 7th edition. Commonwealth Mycological Institute, Kew, 1983.
**Kendrick, B.** *The Fifth Kingdom.* Mycological Publications, Ontario, 1985.
**Oldridge, S G, Pegler, D N & Spooner, B M.** *Wild Mushroom and Toadstool Poisoning.* Royal Botanic Gardens, Kew, 1989.
**Webster, J.** *Introduction to Fungi.* Cambridge University Press, London, 1970.

## North America

**Gilbertson, R L & Ryvarden, L.** *North American Polypores: Abortiporus – Lindtneria,* Volume 1; *Megasporoporia – Wrightoporia,* Volume 2. Fungiflora, Oslo, 1986 & 1987.
**Lincoff, G H.** *The Audubon Society Field Guide to North American Mushrooms.* Alfred A Knopf, New York, 1981.
**Miller, O K Jr.** *Mushrooms of North America.* Dutton & Co, New York, 1973.
**Miller, O K Jr & Miller, H H.** *Gasteromycetes. Morphological and Development Features.* Mad River Press, Eureka, 1988.
**Phillips, R.** *Mushrooms of North America.* Little, Brown & Company, Boston, 1991.
**Smith, A H.** *A Field Guide to Western Mushrooms.* Ann Arbor: University of Michigan, 1975.
**Smith, A H.** *The Mushroom Hunter's Field Guide,* 2nd edition. Ann Arbor: University of Michigan, 1963.
**Smith Weber, N & Smith, A H.** *A Field Guide to Southern Mushrooms.* Ann Arbor: University of Michigan, 1985.

## Europe

**Bon, M.** *The Mushrooms and Toadstools of Britain and North-western Europe.* Hodder & Stoughton, London, 1987.
**Breitenbach, J & Kränzlin, F.** *Fungi of Switzerland.* Volume 1. *Ascomycetes.* Verlag Mykologia, Lucerne, 1984.
**Breitenbach, J & Kränzlin, F.** *Fungi of Switzerland.* Volume 2. *Non-gilled Fungi.* Verlag Mykologia, Lucerne, 1986.
**Breitenbach, J & Kränzlin, F.** *Fungi of Switzerland,* Volume 3, *Boletes and Agarics, first part.* Verlag Mykologia, Lucerne, 1991.
**Buczacki, S.** *Fungi of Britain and Europe.* Collins New Generation Guide, London, 1989.
**Dennis, R W G.** *British Ascomycetes,* 3rd edition. J Cramer, Vaduz, 1978.
**Ellis, M B & Ellis, J. P.** *Fungi Without Gills – An Identification Handbook.* Chapman & Hall, London, 1990.
**Eriksson, J, Hjortstam, K, Larssen, K-H & Ryvarden, L.** *The Corticiaceae of Northern Europe.* Volumes 1–8. Fungiflora, Oslo, 1973–87.
**Holden, M.** *Guide to the Literature for the Identification of British Fungi,* 4th edition. British Mycological Society, 1982.
**Jülich, W.** *Die Nichblätterpilze, Gallertpilze und Bauchpilze.* Kleine Kryptogamenflora, Volume IIb/1. Verlag Fischer, Stuttgart, 1984.
**Lange, M & Hora, B A.** *Collins Guide to Mushrooms & Toadstools.* Collins, London, 1963.
**Moser, M.** *Keys to Agarics and Boleti (Polyporales, Boletales, Agaricales, Russulales),* English edition. Roger Phillips, London, 1983.
**Pegler, D N.** *Field Guide to the Mushrooms and Toadstools.* Kingfisher Books, London, 1990.
**Phillips, R.** *Mushrooms and other Fungi of Great Britain and Europe.* Pan Books, London, 1981.
**Ryvarden, L.** *The Polyporaceae of North Europe.* Volumes 1–2. Fungiflora, Oslo, 1976–78.


# Glossary

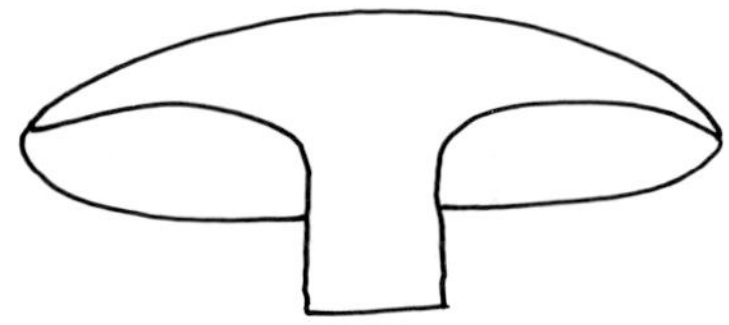

**ADNATE** (of gills, tubes) – broadly attached to stem apex

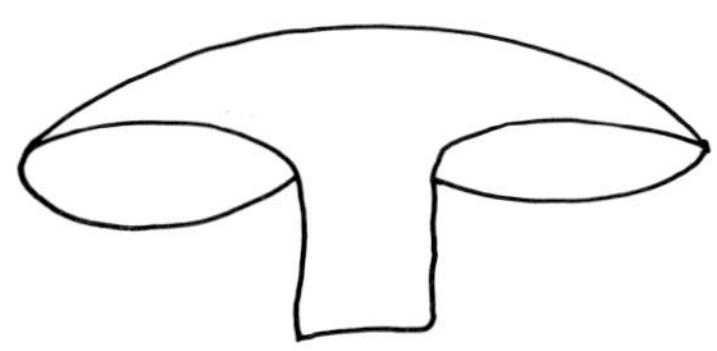

**ADNEXED** (of gills, tubes) – narrowly attached to stem apex
**ASCUS** (pl. asci) – the spore-producing "cell" of an Ascomycete, typically forming eight spores internally
**BASIDIUM** (pl. basidia) – the spore-producing "cell" of a Basidiomycete,

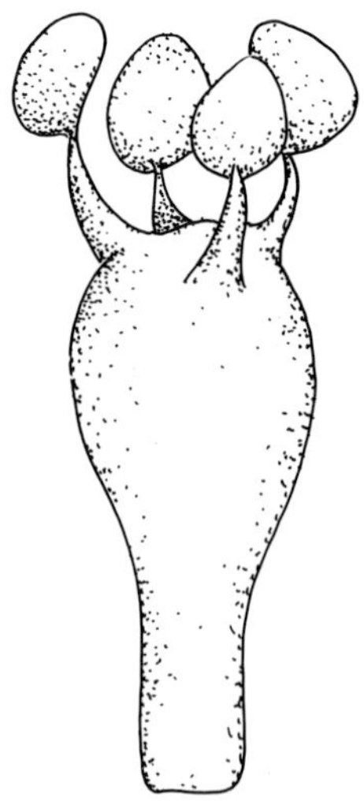

usually club-shaped and forming four spores
**CAMPANULATE** (of cap) – bell-shaped
**CAPITATE** – with a swollen apex

**CONVEX** (of cap) – broadly rounded upwards in outline

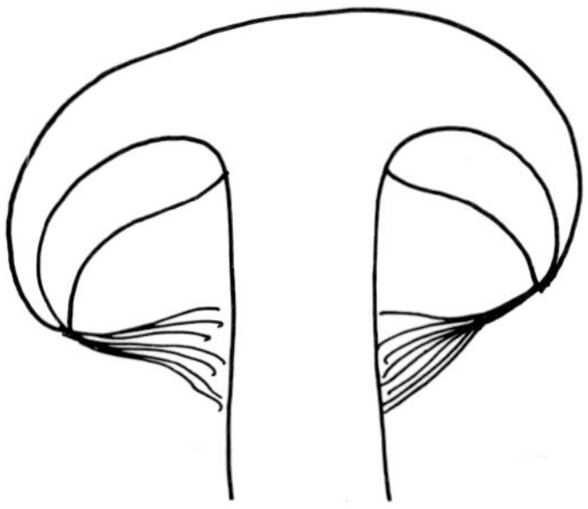

**CORTINA** (of mushrooms) – a cobweb-like veil initially connecting the cap margin to stem

**CRENULATE** – minutely toothed

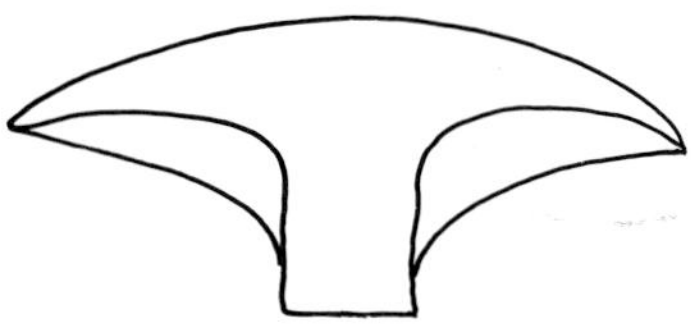

**DECURRENT** (of gills, tubes) – broad attachment which extends down the stem

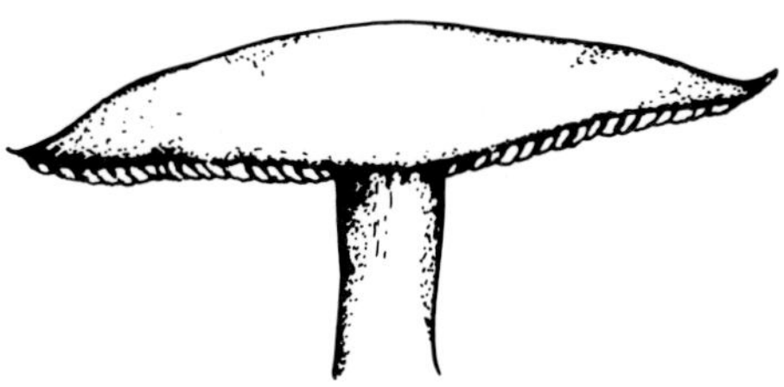

**EXPANDED** (of cap) – opening fully at maturity

**FERTILE SURFACE** – spore-producing surface

**FIBRIL** – a small fibre

**FIBRILLOSE** – bearing fibrils

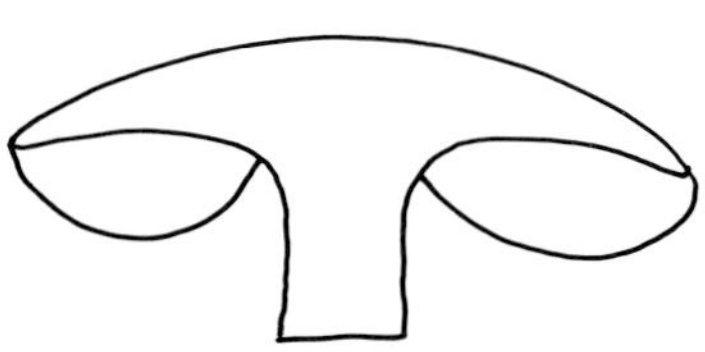

**FREE** (of gills, tubes) – not reaching the stem apex

**FRUITBODY** – the structure which supports the spore-producing layers of a fungus, visible to the naked eye (eg mushroom)

**GLEBA** – the internal, spore-producing tissue of a Gasteromycete

**HYPHA** (pl. hyphae) – an individual thread-like unit which makes up the mycelium and the fruitbody

**LAMELLA** (pl. lamellae) – a gill of a mushroom

**MERULIOID** (of fertile surface) – wrinkled and folded (*Merulius-like)*

**MYCELIUM** – a mass of hyphae, usually embedded within the substrate

**OBTUSE** (of cap margin) – thick and rounded

**PAPILLATE** (of cap) – with a prominent central projection

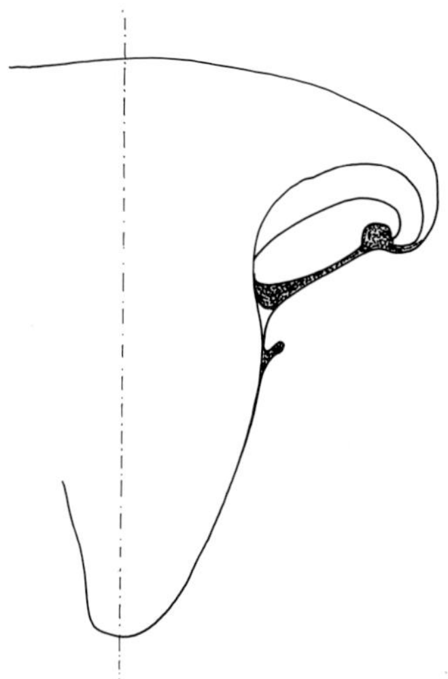

**PARTIAL VEIL** – a protective layer of certain young mushroom fruitbodies

**PERIDIOLE** – packet of spores in Bird's Nest Fungi

**PERITHECIUM** – the flask-shaped fruitbody of a Pyrenomycete (Ascomycetes)

**PILEUS** – the cap of a mushroom or bracket-fungus

**RHIZOMORPH** – a cord-like strand of mycelium, resembling a "root"

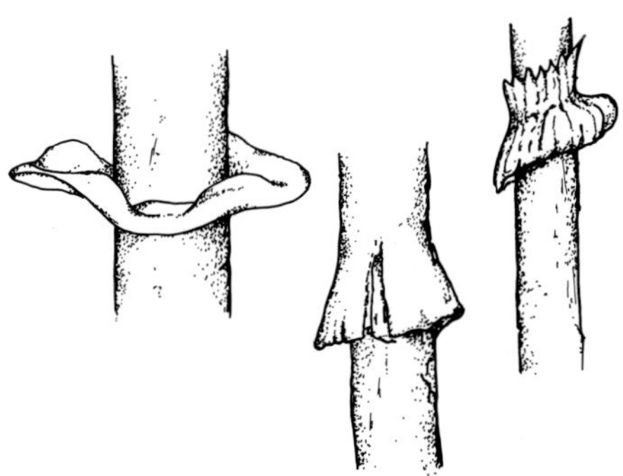

**RING** – the membranous remains of the partial veil, attached to the stem

**SCLEROTIUM** – a sterile mass of closely packed hyphae, acting as a resting stage

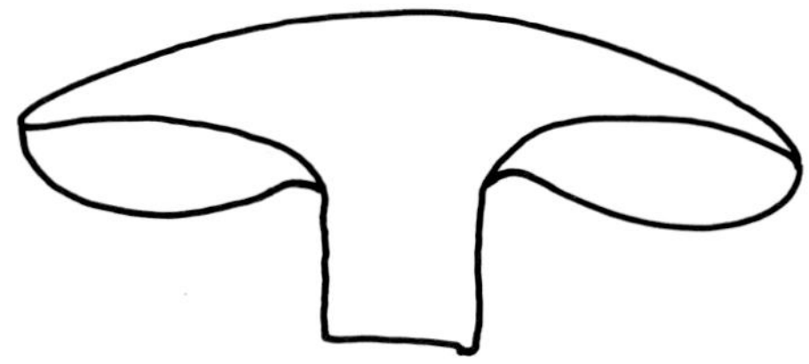

**SINUATE** (of gills, tubes) – curving upwards just before the point of attachment to the stem

**SPORE** – the reproductive, one-celled unit of a fungus; hence ascospore (in Ascomycetes) and basidiospore (in Basidiomycetes)

**SQUARROSE** (of cap) – with erect scales

**STIPE** – the stem of a mushroom

**STROMA** – the sterile flesh of an Ascomycete

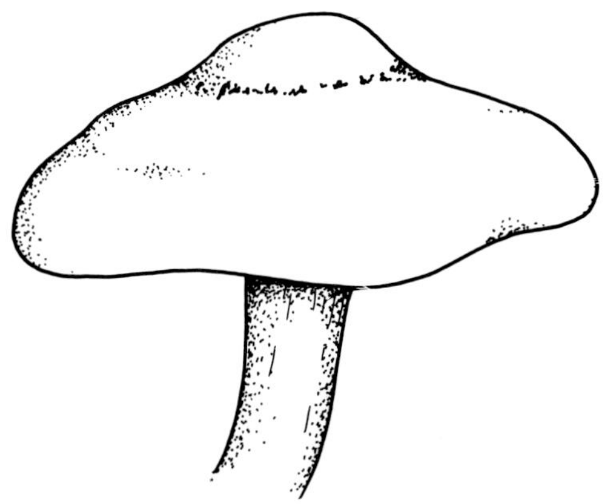

**UMBONATE** (of cap) – having a raised area (or umbo) at the centre

**VEIL** – a protective layer covering the button-stage of mushrooms

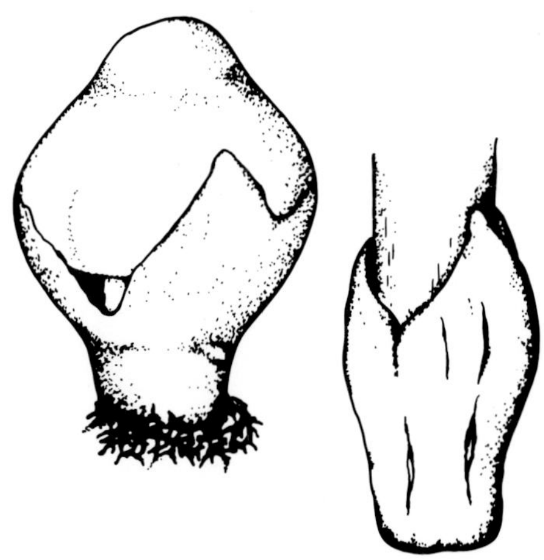

**VOLVA** – a cup-like structure at the stem base, derived from the veil

# INDEX

## OF LATIN AND ENGLISH NAMES IN THE DIRECTORY


*Agaricus* sp. 34, 78, 86
*Agaricus arvensis* 81
*Agaricus augustus* 39
*Agaricus bisporus* 16, 81
*Agaricus bitorquis* 81
*Agaricus campestris* 81
*Agaricus excellens* 39
*Agaricus haemorrhoidarius* 39
*Agaricus macrosporus* 39
*Agaricus placomyces* 82
*Agaricus silvaticus* 39
*Agaricus silvicola* 39
*Agaricus xanthodermus* 16, 81, 82
*Agrocybe cylindrica* 103
*Agrocybe praecox* 84
*Agrocybe semiorbicularis* 84
*Albatrellus confluens* 88
*Albatrellus ovinus* 88
Alcohol Ink-cap 82
Alderwood Polypore 118
*Aleuria* sp. 10, 14
*Aleuria aurantia* 91
*Amanita* sp. 8, 13, 81, 102, 135
*Amanita citrina* 32
*Amanita citrina* var. *alba* 34
*Amanita excelsa* 32, 33
*Amanita flavorubescens* 34
*Amanita formosa* 33
*Amanita fulva* 33
*Amanita junquillea* 32
*Amanita muscaria* 16, 33
*Amanita ovoidea* 34
*Amanita pantherina* 16, 32, 33, 34
*Amanita phalloides* 16, 32, 34
*Amanita regalis* 33
*Amanita rubescens* 16, 32, 33, 34
*Amanita vaginata* 16, 32
*Amanita virosa* 34
Amethyst Deceiver 21
Angel's Wings 94
Anise-scented Clitocybe 24
Arched Earthstar 65
*Armillaria mellea* 95, 105
*Armillaria ostoyae* 95
*Armillaria tabescense* 95
Aromatic Milk-cap 52
Artist's Fungus 116
*Astraeus hygrometricus* 88
*Aurantioporus croceus* 11
*Auricularia* sp. 10
*Auricularia auricula-judae* 124
*Auricularia mesenterica* 124
*Auricularia polytricha* 10
*Auriscalpium vulgare* 107
Autumn Chanterelle 19

*Baeospora myosura* 99
Barometer Earthstar 88
Bay Boletus 58
Beaked Earthstar 65
Bearded Tooth Fungus 108
Beefsteak Fungus 117
Birch Cushion 126
Birch Lenzites 114
Birch Polypore 114
Bitter Boletus 59
*Bjerkandera adusta* 110
*Bjerkandera fumosa* 110
Black Bulgar 125
Black Jelly Drops 125
Black Jelly Roll 123
Black Milk-drop Mycena 30
Black Morel 90
Black Trumpet 19
Black Witches' Butter 123
Blackening Lyophyllum 25
Blackening Meadow Cap 73
Blackening Russula 53
Blackish Purple Russula 56
Black-staining Polypore 113
Bladder Cup 136
Bleeding Conifer Parchment 121
Bleeding Mycena 101
Blewit 22
Blood-red Cortinarius 49
Blood-stained Bracket 111
Blue and Yellow Russula 55
Blue-cheese Polypore 115
Blue-green Clitocybe 24
Blue-toothed Leptonia 78
Bluing Bolete 60
Bluing Psilocybe 85
Blusher, The 34
Bog Beacon 130
Bog-moss Helmet Cap 129
*Bolbitius vitellinus* 84
*Boletus aestivalis* 58
*Boletus dupainii* 58
*Boletus edulis* 58, 59
*Boletus erythropus* 58
*Boletus luridus* 58
*Boletus satanas* 16, 59
*Boletus satanoides* 59
*Bondarzewia berkeleyi* 113
Bonnet Mycena 100
Bossed Lepiota 80
*Bovista* sp. 9
*Bovista nigrescens* 89
*Bovista plumbea* 89
Bracelet Cortinarius 47
Branched Oyster Mushroom 94
Broad-gilled Agaric 98
Brown Birch Bolete 59
Brown Cone Cap 84
Brown Funnel Polypore 64
Brown Hay Cap 83
Brown Pluteus 101
Brown Roll-rim 56
Buff Meadow Cap 73
*Bulgaria inquinans* 123, 125
Burnt Tricholoma 26

Cage Fungus 66
*Calocera* sp. 10
*Calocera viscosa* 88, 123
*Calocybe carnea* 74
*Calocybe gambosa* 75
*Calocybe persicolor* 74
*Calvatia* sp. 9
*Calvatia cyathiformis* 89
*Calvatia excipuliformis* 65, 89
*Calvatia utriformis* 89
*Camarophyllus colemannianus* 73
*Camarophyllus niveus* 73
*Camarophyllus pratensis* 73
*Camarophyllus russocoriaceus* 73
*Camarophyllus virgineus* 73
Campanulate Naucoria 129
Candle Snuff Fungus 126
*Cantharellus cibarius* 19, 56
*Cantharelluis lutescens* 19
*Cantharellus tubaeformis* 19
Carbon Antlers 126
Carbon Balls 125
Carpet-pin Mycena 76
Cauliflower Fungus 107
Cellar Fungus 119
Cep 58
*Chaliciporus piperatus* 86
Changeable Melanoleuca 31
Changing Pholiota 103
Chanterelle 19, 56
Charcoal Burner 55
Charcoal Collybia 132
Charcoal Pholiota 132
Chater's Vermilion Cup 91
Chestnut Bolete 60
Chestnut Parasol 37
Chicken of the Woods 113
*Chlorophyllum molybdites* 16, 38, 80
*Chondrostereum purpureum* 121, 124
*Chroogomphus rutilus* 61
*Chroogomphus vinicolor* 61
Cinnabar-red Polypore 115
Cinnamon Polypore 64
Citron Amanita 32
*Clathrus* sp. 10
*Clathrus archeri* 66
*Clathrus ruber* 66
*Clavaria* sp. 123
*Clavaria acuta* 87
*Clavaria vermicularis* 87
*Clavariadelphus fistulosus* 62
*Clavariadelphus pistillaris* 62
*Clavariadelphus truncatus* 62
*Clavulina cinerea* 62
*Clavulina cristata* 62
*Clavulina rugosa* 63
*Clavulinopsis corniculata* 88
*Clavulinopsis fusiformis* 87
*Clavulinopsis helvola* 87
*Clavulinopsis luteoalba* 87
Clayey Hebeloma 43
*Clitocybe* sp. 12, 16, 76
*Clitocybe clavipes* 22
*Clitocybe dealbata* 36, 75
*Clitocybe dicolor* 24
*Clitocybe fragrans* 24
*Clitocybe geotropa* 23
*Clitocybe infundibuliformis* 23
*Clitocybe nebularis* 23
*Clitocybe odora* 24
*Clitocybe phyllophila* 24
*Clitocybe rivulosa* 75
*Clitocybe squamulosa* 23
*Clitocybe suaveolens* 24
*Clitocybe vibecina* 36
*Clitopilus prunulus* 36
Clouded Clitocybe 23
Club 62
Club-footed Clitocybe 22
Clustered Coral Fungus 63
Clustered Grey Gill 75
Clustered Psathyrella 103
Clustered Tough Shank 28
Cockleshell Lentinus 95
Collared Earthstar 65
*Collybia* sp. 12
*Collybia butyracea* 28, 97
*Collybia confluens* 28
*Collybia dryophila* 29
*Collybia fusipes* 97
*Collybia maculata* 29
*Collybia peronata* 28, 29
*Collybia tuberosa* 138
*Coltricia cinnamomea* 64
*Coltricia perennis* 64
*Coltricia* sp. 117
Common Bird's Nest 123
Common Earthball 66
Common Fiber Vase 64
Common Funnel Cap 23
Common Ink-cap 82
Common Morel 90
Common Mycena 100
Common Psathyrella 41
Common Puffball 66
Common Scaber Stalk 59
Common Stinkhorn 67
Common Violet-latex Milk-cap 53
Common White Helvella 68
Common White Inocybe 44
Common Yellow Russula 54
Conical Wax Cap 73
Conifer-base Polypore 112
*Coniophora arida* 119
*Coniophora puteana* 119
*Conocybe filaris* 85
*Conocybe tenera* 84
Convoluted Bonfire Cup 133
*Coprinus* sp. 13
*Coprinus acuminatus* 82
*Coprinus atramentarius* 16, 82
*Coprinus comatus* 40, 82
*Coprinus disseminatus* 40
*Coprinus domesticus* 40
*Coprinus micaceus* 40
*Coprinus picaceus* 40, 82
*Coprinus plicatilis* 83
Coral Spine-fungus 108
*Cordyceps* sp. 10, 14
*Cordyceps canadensis* 139
*Cordyceps capitatus* 139
*Cordyceps militaris* 92
*Cordyceps ophioglossoides* 139
*Coriolus* sp. 114
*Coriolus hirsutus* 110
*Coriolus pubescens* 110
*Coriolus veriscolor* 110, 121
*Cortinarius* sp. 14, 45
*Cortinarius alboviolaceus* 21
*Cortinarius amoenolens* 45
*Cortinarius anthracinus* 49
*Cortinarius armillatus* 47
*Cortinarius bolaris* 48
*Cortinarius calochrous* 45
*Cortinarius cinnabarinus* 49
*Cortinarius collinitus* 46
*Cortinarius cotoneus* 47
*Cortinarius elatior* 46
*Cortinarius hemitrichus* 48
*Cortinarius malachius* 48
*Cortinarius orellanus* 47
*Cortinarius paleacus* 48
*Cortinarius palustris* 129
*Cortinarius phoenicius* 49
*Cortinarius pholideus* 47
*Cortinarius pseudosalor* 46
*Cortinarius purpurascens* 46
*Cortinarius speciosissimus* 16, 47
*Cortinarius spilomeus* 48
*Cortinarius subpurpurascens* 46
*Cortinarius torvus* 48
*Cortinarius trivialis* 46
Cow Horn Clavaria 88
Crab Russula 54
Cracked Green Russula 54
Cracking Clitocybe 75
Cramp Balls 125
*Craterellus cornucopioidies* 7, 19
*Craterellus fallax* 19
*Creolophus cirrhatus* 108
*Crepidotus* sp. 100
*Crepidotus calolepis* 106
*Crepidotus mollis* 106
Crested Coral Fungus 62
*Crucibulum laeve* 123
*Crustlike Cup* 132
*Cultivated Mushroom* 81
*Curry-scented Milk-cap* 52
*Cyathus* sp. 9
*Cyathus olla* 122, 123
*Cyathus striatus* 122
*Cystoderma amianthinum* 38
*Cystoderma amianthinum* var. *rugosoreticulatum* 38
*Cystoderma carcharias* 38
*Cystoderma granulosum* 38

*Dacrymyces* sp. 10
*Daedalea quercina* 111
*Daedaleopsis confragosa* 111
*Daedaleopsis confragosa* var. *tricolor* 111
*Daldinia* sp. 14
*Daldinia concentrica* 125
*Daldinia vernicosa* 125
Dark Downy Paxillus 57
Dead Man's Fingers 126
Deadly Conocybe 85
Death Cap 34
Deceiver, The 21
Decorated Mop 96
*Dermocybe sanguinea* 49
*Dermocybe uliginosa* 129
Destroying Angel 34
Devil's Boletus 59
Devil's Fingers 66
Dingy Tricoloma 27
*Discina perlata* 69
*Disciotis venosa* 69
Dog Stinkhorn 67
Dotted Stem Boletus 58
Dry Rot Fungus 119
Dryad's Saddle 109
Dung Mottle-gill 135
Dung Roundhead 136
Dung Smooth Head 136
Dusty Puffball 89
Dye Polypore 117

Early Cup Fungus 136
Earth Fan 64
Elegant Polypore 109
Elegant Stinkhorn 67
Elf Cup 71
Emetic Russula 55
Eno-take 97
*Entoloma aprile* 36
*Entoloma clypeatum* 27, 36
*Entoloma nidorosum* 77
*Entoloma rhodopolium* 77
*Entoloma sinuatum* 16, 23, 36
*Exidia glandulosa* 123, 125
Eyelash Cup 125
Eyelash Fungus 125

Fairy Cake Hebeloma 42
Fairy Ring Champignon 76
False Chanterelle 19, 56
False Death Cap 32
False Morel 67
False Tinder Fungus 118
Fawn Pluteus 101
Fetid Marasmius 97
Field Mushroom 81
Fir Polystictus 110
Fir-cone Hydnum 107
*Fistulina hepatica* 117
Flaky Lepiota 80
*Flammulina velutipes* 16, 97
Flat-footed Clitocybe 22
Flat-topped Agaricus 82
*Flavoscypha* sp. 69
*Falvoscypha cantharella* 70
*Flavoscypha concinna* 70
Fleck-flesh Polypore 118
Fleecy Milk-cap 49, 77
Fluted White Helvella 68
Fly Agaric 33
*Fomes foentarius* 112
*Fomitopsis pinicola* 112
Foxy Orange Cortinarius 47
Fragile Russula 55
Freckle-gilled Gymnopilus 105
Fried Chicken Mushroom 75
Fringed Crumble Cap 41
Fruity Cortinarius 45
Funnel Clitocybe 23

*Galerina* sp. 15
*Galerina praticola* 104
*Galerina sphagnorum* 129
*Galerina tibiicystis* 129
*Galerina unicolor* 103, 104
*Ganoderma* sp. 13, 112

*Ganoderma adspersum* 116
*Ganoderma applanatum* 116
*Ganoderma carnosum* 116
*Ganoderma curtisii* 116
*Ganoderma lucidum* 116
*Ganoderma oregonense* 116
*Ganoderma tsugae* 116
Garland Stropharia 86
Garlic Mushroom 35
*Geastrum* sp. 9, 88
*Geastrum fornicatum* 65
*Geastrum pectinatum* 65
*Geastrum quadrifidum* 65
*Geastrum saccatum* 65
*Geastrum triplex* 65
Gem-studded Puffball 66
*Geoglossum* sp. 92
*Geopora foliacea* 71
*Geopora sumneriana* 71
Geranium-scented Cortinarius 48
Giant Clitocybe 77
Giant Club 62
Giant Polypore 113
Giant Puffball 89
Gilled Bolete 57
Girolle 19
Glistening Ink-cap 40
*Gloeophyllum* sp. 111
*Gloeophyllum abietinum* 114
*Gloeophyllum sepiarium* 114
Golden False Pholiota 49
Golden Granular Pluteus 102
Golden Meadow Club 87
Golden Pholiota 104
Golden Spindles 87
Goldenthread Cordyceps 139
Golden-spotted Waxy Cap 26
*Gomphidius* sp. 8
*Gomphidius glutinosus* 61, 62
*Gomphidius roseus* 62
*Gomphus clavatus* 20
*Gomphus floccosus* 20
Granulated Boletus 60
Greasy Tough Shank 28, 97
Green Earth Tongue 91
Green Stropharia 41
Green-spored Lepiota 80
Gregarius Elf Cap 100
Grey Coral Fungus 62
Grey Mottle-gill 135
Grey Tricholoma 27
Grey-brown Clitocybe 24
*Grifola frondosa* 64, 113
Grisette 32
Grooved Bird's Nest Fungus 122
*Gymnopilus hybridus* 105
*Gymnopilus junonius* 104, 105
*Gymnopilus penetrans* 105
*Gymnopilus sapineus* 105
Gypsy Mushroom 45
*Gyromitra esculenta* 16, 67
*Gyroporus castaneus* 58, 60
*Gyropous cyaneus* 60

Hairy Earth Tongue 92
Hairy Parchment 120
Half Free Morel 91
*Hapalopilus nidulans* 111
*Hebeloma crustuliniforme* 16, 42
*Hebeloma longicaudum* 42
*Hebeloma mesophaeum* 42
*Hebeloma radicosum* 43
*Hebeloma sacchariolens* 43
*Hebeloma sinapizans* 43
Hedgehog Fungus 108
*Helvella* sp. 10
*Helvella acetabulum* 68
*Helvella crispa* 68
*Helvella elastica* 68
*Helvella lacunosa* 68
*Helvella latispora* 68
*Helvella leucomelas* 68
Hemispheric Agrocybe 84
Hemlock Varnish Ganoderma 116
Hen of the Woods 113
Herald of the Winter 20
*Hericium clathroides* 108
*Hericium erinaceum* 108
*Hericium coralloides* 108
*Heterobasidion annosum* 112
*Hirschioporus abietinus* 110
*Hirschioporus fuscoviolaceus* 110
Honey Fungus 95, 105
Horn of Plenty 19
Horse Mushroom 81
Horsehair Toadstool 99
*Hygrocybe citrinovirens* 74
*Hygrocybe coccinea* 74
*Hygrocybe conica* 73
*Hygrocybe conicoides* 73
*Hygrocybe nigrescens* 73
*Hygrocybe psittacina* 74
*Hygrocybe punicea* 74
*Hygrophoropsis aurantiaca* 19, 56
*Hygrophorus chrysodon* 20
*Hygrophorus cossus* 20
*Hygrophorus hypothejus* 20
*Hygrophorus olivaceo-albus* 20
*Hymenochaete rubiginosa* 122
*Hymenochaete tabacina* 122
*Hyphoderma sambuci* 120
*Hypholoma* sp. 15
*Hypholoma capnoides* 105
*Hypholoma elongatum* 128
*Hypholoma fasciculare* 105
*Hypholoma sublateritium* 105
*Hypholoma udum* 128
*Hypoxylon multiforme* 126

*Inocybe fastigiata* 44
*Inocybe geophylla* 44
*Inocybe geophylla* var. *lilacina* 44
*Inocybe jurana* 86
*Inocybe lanuginella* 44
*Inocybe lanuginosa* 44
*Inocybe maculata* 44
*Inocybe napipes* 45
*Inocybe patouillardii* 86
*Inocbye umbrina* 45
*Inonotus dryadeus* 117
*Inonotus dryophilus* 117
*Inonotus radiatus* 118
Ivory Clitocybe 75

Jack O'Lantern 105
Jelly Antler Fungus 88, 123
Jelly Babies 130
Jelly Crepidotus 106
Jelly Rot 119
Jew's Ear 124

King Alfred's Cakes 125
*Kuehneromyces mutabilis* 103, 104

*Laccaria amethystea* 21
*Laccaria laccata* 21
*Laccaria proxima* 21
*Laccaria trullisata* 21
*Lacrymaria polytricha* 83
*Lacrymaria velutina* 83
*Lactarius* sp. 14, 16
*Lactarius azonites* 50
*Lactarius camphoratus* 52
*Lactarius chrysorrheus* 51
*Lactarius decipiens* 51
*Lactarius deliciosus* 50
*Lactarius deterrimus* 50
*Lactarius glyciosmus* 51
*Lactarius helvus* 50
*Lactarius hepaticus* 52
*Lactarius hygrophoroides* 52
*Lactarius maculatus* 53
*Lactarius mairei* 50
*Lactarius piperatus* 49, 77
*Lactarius pterosporus* 50
*Lactarius pubescens* 50
*Lactarius rufus* 52
*Lactarius tabidus* 51
*Lactarius theiogalus* 51
*Lactarius torminosus* 50
*Lactarius turpis* 51
*Lactarius uvidus* 53
*Lactarius vellereus* 49, 53, 77
*Lactarius volemus* 52
*Laetiporus sulphureus* 113
*Langermannia gigantea* 89
*Langermannia* sp. 9
Large Lentinus 108
Large Pine Polyporus 117
Large-spored Mushroom 39
Lawn Puffball 90
Leaf Marasmius 99
*Leccinum aurantiacum* 60
*Leccinum holopus* 59
*Leccinum scabrum* 59
*Leccinum versipelle* 60
*Leccinum vulpinus* 59
Lemon Peel Fungus 69
*Lentinellus cochleatus* 95
*Lentinellus omphalodes* 95
*Lentinula edodes* 16
*Lentinus* sp. 9
*Lentinus lepideus* 108
*Lentinus ponderosus* 108
*Lentinus strigosus* 109
*Lentinus tigrinus* 109
*Lentodium squamulosum* 109
*Lenzites betulina* 114
*Leotia atrovirens* 130
*Leotia lubrica* 130
*Lepiota castanea* 36, 37
*Lepiota clypeolaria* 36
*Lepiota cristata* 36, 79
*Lepiota fulvella* 37
*Lepiota subincarnata* 37
*Lepista gilva* 22
*Lepista inversa* 22
*Lepista nuda* 21, 22
*Lepista saeva* 22
*Leptonia euchroa* 78
*Leptonia lampropus* 78
*Leptonia serrulata* 78
*Leucoagaricus naucinus* 78
*Leucocoprinus birnbaumii* 79
*Leucocoprinus brebissonii* 79
*Leucocoprinus cepaestipes* 79
*Leucocoprinus luteus* 79
*Leucopaxillus giganteus* 23, 77
Liberty Cap 85
Lilac Inocybe 44
Lilac Mycena 31
*Limacella glioderma* 35
*Limacella lenticularis* 35
Ling Chih 116
Liquorice Milk-cap 50
Little Jap Umbrella 83
Little Wheel Toadstool 99
Luminescent Panellus 100
*Lycoperdon* sp. 9, 90
*Lycoperdon foetidum* 66
*Lycoperdon lividum* 122
*Lycoperdon perlatum* 66
*Lycoperdon pyriforme* 122
*Lyophyllum connatum* 24
*Lyophyllum decastes* 75
*Lyophyllum fumatofoetens* 25
*Lyophyllum fumosum* 75
*Lyophyllum loricatum* 75

*Macrolepiota excoriata* 80
*Macrolepiota gracilenta* 38, 80
*Macrolepiota mastoidea* 80
*Macrolepiota procera* 38
*Macrolepiota rhacodes* 38
Magpie Ink-cap 40, 82
Malodorous Lepiota 36
*Marasmiellus candidus* 96
*Marasmiellus ramealis* 96
*Marasmius alliaceus* 35
*Marasmius androsaceus* 99
*Marasmius capillaripes* 99
*Marasmius cohaerens* 35
*Marasmius epiphyllus* 99
*Marasmius oreades* 75, 76
*Marasmius rotula* 99
*Marasmius scorodonius* 35
*Marasmius wynnei* 35
Marginate Galerina 104
Mazegill 111
Mealy Tubaria 106
*Megacollybia platyphylla* 98
*Melanoleuca brevipes* 31
*Melanoleuca cognata* 31
*Melanoleuca grammopodia* 31
*Melanoleuca melaleuca* 31
*Melanophyllum echinatum* 37
*Melastiza chateri* 91
*Meripilus giganteus* 64, 113
*Merulius tremellosus* 119
Mica Cap 40
*Microglossum olivaceum* 92
*Microglossum viride* 92
*Micromphale brassicolens* 97
*Micromphale foetidum* 97
*Micromphale perforans* 99
Milk-drop Mycena 30
Milk-white Polypore 115
Milk-white Russula 53
Miller, The 36
*Mitrula paludosa* 130
Moist Lactarius 53
*Morchella* sp. 10, 14
*Morchella conica* 90
*Morchella elata* 90, 91
*Morchella esculenta* 90
*Morchella hortensis* 90
*Morchella semilibera* 91
Mosaic Puffball 89
Mower's Mushroom 83
Mushroom of Immortality 116
Mustard Yellow Polypore 118
*Mutinus* sp. 10
*Mutinus caninus* 67
*Mutinus elegans* 67
*Mycena alcalina* 77
*Mycena candida* 30
*Mycena crocata* 29
*Mycena epipterygia* 30
*Mycena galericulata* 100
*Mycena galopus* 29, 30
*Mycena galopus* var. *candida* 30
*Mycena haematopus* 29, 101
*Mycena inclinata* 100
*Mycena lepocephala* 77
*Mycena leucogala* 30
*Mycena pelianthina* 31
*Mycena pura* 31
*Mycena sanguinolenta* 29, 101
*Mycena viscosa* 30
*Myxomphalia maura* 132

*Naucoria escharoides* 129
Netted Rhodotus 102
*Nidularia* sp. 9
Nitrous Mycena 77
*Nolenea sericea* 78
Non-inky Corpinus 140
*Nyctalis asterophora* 138
*Nyctalis parasitica* 138

Oak Parchment 120
Oak-loving Collybia 28
Ochre Jelly Club 130
Old Man of the Woods 61
*Omphalina ericetorum* 76
*Omphalina rustica* 76
*Omphalotus olearius* 105
Onion-stalked Lepiota 79
Orange Birch Bolete 60
Orange Moss Agaric 76
Orange Peel Fungus 91
Orange Pholiota 104, 105
Orange Pluteus 102
Orange-brown Milk-cap 52
*Otidea onotica* 69
*Oudemansiella mucida* 98
*Oudemansiella pudens* 98
*Oudemansiella radicata* 98
Ox-eye Lepiota 79
Oyster Mushroom 94

Pale Paxillus 106
*Panaeolina foenisecii* 83
*Panaeolus* sp. 12, 83
*Panaeolus campanulatus* 135
*Panaeolus phalaenarum* 135
*Panaeolus semiovatus* 135
*Panaeolus sphinctrinus* 135
*Panellus stipticus* 100
Panther, The 33
Parasitic Bolete 139
Parasitic Volvaria 138
Parasol Mushroom 38
Parrot Toadstool 74
Pavement Mushroom 81
*Paxillus* sp. 8
*Paxillus atrotomentosus* 57
*Paxillus filamentosus* 56
*Paxillus involutus* 16, 56
*Paxillus panuoides* 106
*Paxillus rubicundulus* 56
Peaked Inocybe 44
Pear-shaped Puffball 122
Peat Sulphur Cap 128
*Peniophora quercina* 120
*Peniophora rufomarginata* 120
Penny Bun Boletus 58
Peppery Bolete 86
Peppery Milk-cap 49, 77
Pestle-shaped Coral 62
Pestle-shaped Puffball 65
*Peziza* sp. 10, 14, 69
*Peziza badia* 70
*Peziza badioconfusa* 70
*Peziza proteana* var. *sparassoides* 133
*Peziza pseudoviolacea* 133
*Peziza vesiculosa* 15, 136
*Peziza violacea* 133
*Phaeolepiota aurea* 49
*Phaeolus schweinitzii* 117
*Phallus* sp. 10
*Phallus impudicus* 67
*Phallus ravenelii* 67
*Phellinus gilvus* 118
*Phellinus igniarius* 112, 118
*Phellinus nigricans* 118
*Phellinus robustus* 118
*Phellinus tremulae* 118
*Phlebia radiata* 118
*Phlebia rufa* 118
*Pholiota adiposa* 104
*Pholiota aurivella* 104
*Pholiota highlandensis* 132
*Pholiota myosotis* 128
*Pholiota spumosa* 132
*Pholiota squarrosa* 104, 105
*Phylloporus rhodoxanthus* 57
*Phyllotopsis nidulans* 106
Pick-a-back Toadstool 138
Pigskin Poison Puffball 66
Pig's Ear Gomphus 20
Pig's Ears 70
Pineapple Pholiota 104
Pine-cone Tooth Fungus 107
Pine Fire Fungus 132
Pine Hebeloma 42
Pine Spike Cap 61
Pink Calocybe 74
Pink Crown 70
Pink-fringed Milk-cap 50
Pink Mycena 31
Pinkish Parasol 37
Pinwheel Marasmius 99
*Piptoporus betulinus* 114
Platterful Mushroom 98
*Pleurocybella porrigens* 94
*Pleurotus cornucopiae* 94
*Pleurotus dryinus* 94
*Pleurotus ostreatus* 16, 94
*Pluteus* sp. 135
*Pluteus atromarginatus* 101
*Pluteus aurantiorugosus* 102
*Pluteus cervinus* 101
*Pluteus leoninus* 102
*Pluteus lutescens* 102
*Pluteus magnus* 101
*Pluteus podospileus* 101
*Pluteus umbrosus* 101
Poached Egg Fungus 98
Poison Paxillus 56
Poison Pie 42
*Polyporus badius* 109
*Polyporus brumalis* 109
*Polyporus melanopus* 109
*Polyporus nummularius* 109
*Polyporus squamosus* 109
*Polyporus tuberaster* 109
*Polyporus umbellatus* 64, 113
*Polyporus varius* 109
Powder Cap 138
Poznan Cortinarius 147
Prince, The 39
*Psathyrella candolleana* 41
*Psathyrella gracilis* 41, 85
*Psathyrella hydrophila* 103
*Psathyrella pennata* 41
*Psathyrella spadicea* 103
*Psathyrella squamosa* 41
*Psilocybe* sp. 16
*Psilocybe coprophila* 136
*Psilocybe cyanescens* 85
*Psilocybe fimetaria* 85
*Psilocybe semilanceata* 85
*Psilocybe squamosa* 42
*Psilocybe subcoprophila* 136
*Psilocybe thrausta* 42
*Pulcherricium caeruleum* 120
Pungent Cystoderma 38
Purple and Yellow Agaric 96
Purple Cortinarius 46
Purple-tipped Coral Fungus 63
*Pycnoporus cinnabarinus* 115
*Pycnoporus sanguineus* 115

Radiating Phlebia 118
*Ramaria* sp. 8
*Ramaria aurea* 63
*Ramaria botrytis* 63
*Ramaria formosa* 63
*Ramaria ochraceovirens* 107
*Ramaria stricta* 107
Razor-strop Fungus 114
Red Inocybe 86
Red-branded Cortinarius 47
Red-brown Crust 122
Red-cracked Boletus 57

Red-dappled Cortinarius 48
Red-gilled Agaric 37
Red-hot Milk-cap 52
Red-rimmed Bracket 112
*Rhizina undulata* 132
*Rhodotus palmatus* 102
Ribbed-stalk Cup 68
*Rickenella fibula* 76
*Rickenella fibula* var. *swartzii* 76
Roman Shield Entoloma 36
Root Fomes 112
Rooting Pholiota 43
Rooting Shank 98
Rose-gilled Grisette 102, 135
Rosy Entoloma 77
Rosy Spike Cap 62
Round Stropharia 136
Rounded Earthstar 65
*Rozites caperata* 45
Rufous Milk-cap 52
Russet Tough Shank 28
Russet-scaly Tricholoma 25
*Russula* sp. 14, 16
*Russula adusta* 53
*Russula aeruginosa* 54
*Russula aquosa* 55
*Russula atropurpurea* 56
*Russula betularum* 55
*Russula claroflava* 54
*Russula crustosa* 54
*Russula cyanoxantha* 54, 55
*Russula delica* 53
*Russula densifolia* 53
*Russula emetica* 55
*Russula fellea* 54
*Russula fragilis* 55, 56
*Russula grisea* 55
*Russula lutea* 54
*Russula mairei* 55
*Russula nigricans* 53
*Russula ochroleuca* 54
*Russula solaris* 54
*Russula virescens* 54
*Russula xerampelina* 54

Saw-gilled Leptonia 78
Saffron Milk-cap 50
Saffron Parasol 38
Salmon Wax Cap 73
*Sarcosphaera crassa* 70
Scaly Cortinarius 47
Scaly Crumble Cap 41
Scaly Lentinus 108
Scaly Psilocybe 42
Scaly Wood Mushroom 39
Scarlet Caterpillar Fungus 92
Scarlet Hood 74
*Schizophyllum commune* 95
*Scleroderma* sp. 9
*Scleroderma citrinum* 66
*Scleroderma geaster* 66
Scurfy Cortinarius 48
*Scutellinia* sp. 14
*Scutellinia scutellata* 125
Scented Hebeloma 43
Separating Tricholoma 26
*Serpula himantioides* 119
*Serpula lacrymans* 118
Shaggy Ink-cap 82
Shaggy Parasol 38
Shaggy Pholiota 104, 105
Sheathed Cortinarius 48
Sheep Polypore 88
Shining Ganoderma 116
Shiny Cobweb Cap 46
Silky Grey Tricholoma 27
Silky Nolanea 78
Silky Volvaria 102
Silver Leaf Fungus 121
Slender Lepiota 80
Slender Psathyrella 85
Slender Truffle Club 139
Slimy Beech Caps 98
Slimy Spike Cap 61
Slimy Veil Limacella 35
Slimy-banded Cortinarius 46
Slippery Jack 61
Small Bleeding Mycena 101
Smeared Cortinarius 46
Smoky Polypore 110
Smooth Lepiota 78
Smooth Thimble Cap 69
Smooth Volvariella 135
Smooth-stalked Helvella 68
Snowy Meadow Cap 73
Soft Slipper Toadstool 106
Southern Poplar Mushroom 103
*Sparassis* sp. 133
*Sparassis crispa* 107
*Sparassis laminosa* 107
Sphagnum-bog Galerina 129
Spindle Shank 97
Splash Cups 122
Split Gill 95
Spotted Tough Shank
Spring Agaric 84
Spring Agricus 81
Spring Pine Cone Toadstool 99
Spring Tricholoma 31
St George's Mushroom 75
Stainer, The 29
Stalkless Paxillus 106
*Stereum gausapatum* 121
*Stereum hirsutum* 121
*Stereum insignitum* 121
*Stereum rugosum* 121
*Stereum sanguinolentum* 121
*Stereum subtomentosum* 121
Stinking Parasol 36, 79
Straight-branched Coral 107
Straw-colored Fiber Head 44
*Strobilomyces strobilaceus* 61
*Strobilurus esculentus* 99
*Strobilurus tenacellus* 99
Strong-scented Entoloma 77
*Stropharia aeruginosa* 41
*Stropharia albocyanea* 41
*Stropharia caerulea* 41
*Stropharia coronilla* 86
*Stropharia semiglobata* 136
Stump Mycena 77
Stump Puffball 122
Styptic Fungus 100
*Suillus collinitus* 60
*Suillus granulatus* 60
*Suillus grevillei* 61
*Suillus luteus* 61
Sulphur Polypore 113
Sulphur Tricholoma 26
Sulphur Tuft 105
Sumner's Earthcup 71
Swamp Dermocybe 129
Swamp Grey Gill 128
Swamp Sulphur Cap 128
Sweet Lactarius 51
Sweetbread Mushroom 36

Tall Amanita 32
*Tarzetta catinus* 71
*Tarzetta cupularis* 71
Tawny Funnel Cap 22
Tawny Grisette 33
Tender Nesting Polypore 111
*Tephrocybe ambusta* 132
*Tephrocybe atrata* 132
*Tephrocybe palustre* 128
*Thelephora anthocephala* 64
*Thelephora terrestris* 64
Thick-maze Oak Polypore 111
Thin-maze Flat Polypore 111
Thread Cone Cap 85
Tiger-spot Lentinus 109
Tinder Fungus 112
Tree Ear 124
Tree Volvaria 102
Trembling Merulius 119
*Tremella* sp. 10
*Tremella delica* 10
*Tremella mesenterica* 124
*Trichologlossum hirsutum* 92
*Tricholoma* sp. 12, 16, 98
*Tricholoma albobrunneum* 25, 26
*Tricholoma equestre* 26
*Tricholoma fulvum* 25
*Tricholoma imbricatum* 25
*Tricholoma portentosum* 27
*Tricholoma scalpturatum* 27
*Tricholoma sejunctum* 26
*Tricholoma sulphureum* 26
*Tricholoma terreum* 27
*Tricholoma ustale* 26
*Tricholoma ustaloides* 26
*Tricholoma vaccinum* 25
*Tricholoma viridilutescens* 26
*Tricholoma potentosum* 27
*Tricholomopsis decora* 96
*Tricholomopsis rutlinas* 96
*Tricholomopsis sulphuroides* 96
Tripe Fungus 124
Trooping Cordyceps 92
Trooping Crumble Caps 40
Trumpet Chanterelle 19
Trumpet Clitocybe 23
*Tubaria conspersa* 106
*Tubaria furfuracea* 106
*Tubaria hiemalis* 106
*Tuber* sp. 14
Tufted Collybia 28
Turban Fungus 67
Turkey-tail 110
Turnip-foot Inocybe 45
Twig Marasmius 96
Two-tone Pholiota 103
*Tylopilus felleus* 59
*Tyromyces caesius* 115
*Tyromyces chioneus* 115
*Tyromyces lacteus* 115
*Tyromyces stipticus* 115
*Tyromyces subcaesius* 115

Ugly Milk-cap 51
Umbrella Navel Cap 76
Umbrella Polypore 64, 113

Varicoloured Bracket 110
Variegated Mop 96
*Vascellum pratense* 90
Veiled Oyster 94
Vein Crust 118
Veined Cup 69
Velvet Blue Spread 120
Velvet Shank 97
Velvet-footed Paxillus 57
Verdigris Agaric 41
*Verpa bohemica* 69
*Verpa conica* 69
Violet Bonfire Cup 133
Voluminous-latex Milk-cap 52
*Volvariella bombycina* 102
*Volvariella speciosa* 102, 135
*Volvariella surrecta* 138
*Volvariella volvacea* 16
Warted Oak Polypore 117
Watery Hypholoma 103
Weeping Polypore 117
Weeping Widow 83
Wet Rot Fungus 119
White Crust 120
White Fiber Head 44
White Lyophyllum 24
White Marasmius 96
White Spindles 87
White Worm Coral Fungus 8
White-cheese Polypore 115
White-egg Bird's Nest 123
Witches' Butter 124
Witch's Hat 73
Wood Blewit 21
Wood Woolly Foot 29
Woolly Inocybe 44
Woolly Milk-cap 50
Wrinckled Club 63
Wynne's Marasmius 35

*Xerocomus badius* 58
*Xerocomus chrysenteron* 57
*Xerocomus parasiticus* 139
*Xerocomus subtomentosus* 57
*Xylaria* sp. 10, 14
*Xylaria carpophila* 126
*Xylaria hypoxylon* 126
*Xylaria longipes* 126
*Xylaria polymorpha* 126

Yellow Brain Fungus 124
Yellow Cottony Agaric 79
Yellow Cow-pat Toadstool 8
Yellow Downy Hygrophorus 20
Yellow Hare's Ear 70
Yellow Milk-cap 51
Yellow Morel 90
Yellow Oyster Mop 96
Yellow Pluteus 102
Yellow Stereum 121
Yellow-tipped Coral Fungus
Yellow Tuning Fork 123
Yellow-brown Tricholoma 25
Yellow-foot Agaricus 82
Yellow-red Gill Polypore 114
Yellow-staining Mushroom 8 82
Yellow-staining Tricholoma 2
Yellow-stemmed Mycena 30

# PICTURE CREDITS

Numerals refer to page numbers; L = Left; C = Center; R = Right; T = Top; B = Bottom.

**A W Brand:** 8 TL; 14; 18; 19 R; 21 L; 28 C; 29 C, R; 30 C, R; 32 L, C; 38 L, R; 41 R; 42 C; 51 L; 52 L; 54 L; 55 R; 56 C; 58 R; 59 R; 64 L; 65 C; 66 C; 67 C; 69 C; 77 L, C; 79 C; 81 L, R; 85 L, C; 87 B; 92 L, C; 95 L, R; 96 C, R; 98 C, R; 99 C; 100 L; 101 R; 103 L; 104 C; 109 C; 110 R; 111 R; 122 C; 124 C; 125 R; 135 R; 137; 138 C; 139 C. **Gordon Dickson:** 8 BR; 9 R; 15; 19 L; 20 R; 24 C; 25 L, R; 26 C, R; 27 L; 30 L; 31 L; 36 R; 37 L; 39 L; 41 C; 42 R; 43 R; 44 L, C, R; 46 C; 47 L, C, R; 48 C; 49 L; 50 R; 51 R; 53 L; 58 L; 59 L, C; 60 R; 62 L, C; 66 R; 68 L, R; 69 R; 73 L; 75 L, R; 76 L; 79 L; 80 C; 81 C; 82 R; 83 L; 85 R; 86 L; 87 C, T; 88 L; 89 C; 91 C, R; 94 L, R; 100 R; 102 C; 103 R; 105 C; 107 C; 108 L, R; 109 L; 110 L; 111 C; 112 L; 113 C; 117 C; 118 C; 119 L; 120 R; 121 T; 122 R; 127; 128 L; 129 L; 130 L, C; 133 T; 135 C. **Margaret Holden:** 20 L; 88 R; 128 R. **Thomas Laessøe:** 6; 7; 9 L; 24 R; 28 R; 34 C; 35 L; 36 L; 38 C; 60 L; 70 R; 73 R; 74 C; 83 R; 84 C; 89 L; 92 R; 97 L; 101 L; 105 R; 106 L; 112 C, R; 114 C; 117 L; 123 C, R; 124 R; 126 L, C, R; 129 C; 135 L; 136 R; 138 R. **Nick Legon:** 21 R; 26 L; 39 R; 65 L; 70 C; 74 L; 104 L; 118 R; 120 L. **Alan Outen:** 8 TR; 10 B, T; 20 C; 22 L, C, R; 23 L, C; 24 L; 25 C; 27 C; 31 R; 32 R; 33 C; 34 R; 35 C; 36 C; 37 C, R; 40 R; 41 L; 42 L; 43 L; 45 C, R; 46 L, R; 48 R; 49 R; 51 C; 52 C; 53 R; 55 L; 56 R; 57 R; 58 C; 61 C, R; 62 R; 64 R; 66 L; 67 L; 68 C; 69 L; 70 L; 76 R; 77 R; 78 L, C; 79 R; 82 L, C; 83 C; 84 R; 86 R; 89 R; 90 L, R; 91 L; 93; 97 R; 98 L; 99 R; 100 C; 101 C; 102 L, R; 103 C; 105 L; 106 C; 108 C; 109 R; 111 L; 113 L, R; 114 L, R; 116 C; 119 C, R; 120 C; 121 B, C; 122 L; 123 L; 124 L; 125 L, C; 128 C; 129 R; 131; 132 L, C, R; 133 C; 134; 136 L, C; 138 L; 139 L. **David Pegler:** 80 R. **Jens H Petersen:** 8 BL; 17; 19 C; 21 C; 27 R; 28 L; 29 L; 31 C; 33 L, R; 34 L; 39 C; 40 L, C; 43 C; 45 L; 50 C; 52 R; 53 C; 54 C, R; 55 C; 56 L; 57 L, C; 61 L; 63 C, R; 64 C; 71 T; 72; 73 C; 74 R; 75 C; 76 C; 80 L; 84 L; 66 C; 88 C; 90 C; 94 C; 94 C; 95 C; 97 C; 104 R; 106 R; 107 L; 110 C; 115 L, C, R; 116 L, R; 117 R; 118 L. **Brian Spooner:** 10 R; 23 R; 48 L; 49 C; 50 L; 60 C; 63 L; 65 R; 67 R; 71 C; 78 R; 99 L; 107 R.